The Ultimate Quiz Book

Develop Focus and Aptitude with Engaging Quizzes

Dr. Zeenat Hasan Ph.D.

TABLE OF CONTENTS

1. INTRODUCTION 4
2. HISTORY QUIZ 11
3. GEOGRAPHY QUIZ 38
4. MATHS QUIZ 64
5. PHYSICS QUIZ 92
6. BIOLOGY QUIZ 123
7. SPORTS QUIZ 152
8. ANSWERS 166
9. REFERENCES 197
10. GLOSSARY 200

INTRODUCTION

Welcome to **The Ultimate Quiz Book**, a journey into the world of knowledge, curiosity, and mental agility. This book is more than just a collection of questions and answers; it's a comprehensive tool designed to enhance your knowledge, improve your IQ, and sharpen your aptitude, regardless of your age.

Why This Book is Special

In an era where information is abundant and attention spans are fleeting, it's crucial to have resources that not only inform but also engage. This book aims to do just that. By providing a diverse range of topics and thoughtfully crafted questions, it encourages readers to think critically, expand their horizons, and have fun while learning.

Whether you are a student looking to ace your exams, a professional aiming to stay sharp, or simply a curious mind wanting to challenge yourself, this quiz book is tailored for you. The questions vary in difficulty and cover a wide array of subjects, ensuring that there is something for everyone. From history and science to pop culture and sports, every chapter is a new adventure.

Enhancing Knowledge and IQ

Studies have shown that regularly engaging in activities that challenge the brain can significantly improve cognitive functions. This quiz book is designed to do just that. By testing your knowledge across different domains, it promotes memory retention, enhances problem-solving skills, and boosts overall intellectual capacity.

Quizzes are a proven method to increase brain plasticity, the ability of the brain to adapt and change throughout life. This book's diverse range of questions stimulates different parts of the brain, keeping it active and engaged. Regular use can lead to improved critical thinking, better decision-making abilities, and a sharper mind.

Improving Aptitude

Aptitude isn't just about being smart; it's about being able to apply knowledge effectively. This book helps improve your aptitude by presenting questions that require not just rote memory but also application and reasoning. By challenging yourself with these questions, you enhance your analytical skills and ability to connect the dots, both of which are essential in real-life scenarios.

The Author's Journey

Dr. Zeenat Hasan is Bachelor of Engineering in Computer Science , MBA in Marketing and Information Technology and Doctor of Philosophy in Information Technology and Operations Management . She has published 16 research papers in international journals. She has wide experience working in senior Director roles in Telecom and Banking Industry for last 25 years .

With continuous exposure to irrelevant and unrealistic content through various digital channels, playing with reward and penalty feelings in human mind . It has started rusting the minds , leading to lessening of cognitive skills and memory degradation across age groups.

In order to sharpen the mind and build power to focus, challenging the brain on real world topics is the key. The book helps people to improve their aptitude and focus ,remain relevant , sharpen their intelligence by testing and growing their mind .

She has been practicing it for last 25 years , and it has helped her improve her aptitude and focus to solve critical business and technical problems in hand.

Her goal was to create a resource that is not only educational but also timeless and engaging. She meticulously selected topics and crafted questions that are relevant, thought-provoking, and enjoyable.

Dr. Hasan's extensive research involved consulting academic resources, staying up-to-date with current events, and even testing the questions with diverse groups of people to ensure they are both challenging and accessible. Her commitment to excellence shines through every page of this book.

Timeless Engagement

One of the core principles guiding the creation of this book was the need to keep it timeless. The questions are designed to remain relevant and interesting, regardless of when you pick up the book. This timelessness is achieved by focusing on fundamental knowledge and universally engaging topics.

Moreover, the book is structured to cater to a wide age range, from young learners to seasoned adults. The language is clear and accessible, and the difficulty of questions is balanced to provide a rewarding challenge for all.

A Tool for All Ages

The Ultimate Quiz Book is a testament to the idea that learning never stops. It is a valuable resource for educators, parents, students, and anyone who enjoys a good challenge. Its versatility makes it a perfect companion for family gatherings, study sessions, or even solo brain workouts.

In conclusion, this book is more than just a quiz book. It is a gateway to a smarter, more informed, and intellectually vibrant you. Dr. Zeenat Hasan's dedication and expertise have created a tool that promises to make learning an enjoyable and lifelong pursuit. So, flip the page, take a deep breath, and embark on this exciting journey of discovery and growth.

Happy Quizzing!

The Ultimate Quiz Book

HISTORY QUIZ

1. What was the primary writing system of ancient Egypt?

    a) Hieroglyphics

    b) Cuneiform

    c) Latin

    d) Sanskrit

2. Who was the first emperor of China?

    a) Han Wudi

    b) Liu Bang

    c) Qin Shi Huang

    d) Sun Tzu

3. Which ancient civilization built the city of Machu Picchu?

    a) Aztec

    b) Inca

    c) Maya

    d) Olmec

4. What was the Code of Hammurabi?

    a) An ancient Egyptian law code

b) A Roman code of law

c) A Babylonian law code

d) A Greek legal system

5. Who was the king of the Akkadian Empire who created the world's first known empire?

   a) Hammurabi

   b) Sargon of Akkad

   c) Gilgamesh

   d) Ramses II

6. Who is considered the "Father of Western Philosophy"?

   a) Plato

   b) Aristotle

   c) Socrates

   d) Heraclitus

7. Which war was fought between Athens and Sparta?

   a) Persian War

   b) Trojan War

   c) Peloponnesian War

   d) Macedonian War

8. What is the significance of the Battle of Thermopylae?

   a) It marked the end of the Persian Empire

   b) A small Greek force held off a much larger Persian army

c) It was the final battle of the Trojan War

d) It led to the rise of the Roman Empire

9. Who was the first Roman emperor?

   a) Julius Caesar

   b) Augustus Caesar

   c) Nero

   d) Tiberius

10. What was the Pax Romana?

    a) A Roman peace treaty with Greece

    b) A period of peace and stability throughout the Roman Empire

    c) A Roman military conquest

    d) The founding of the Roman Republic

11. What event is considered the start of the Middle Ages?

    a) The fall of the Western Roman Empire

    b) The birth of Charlemagne

    c) The signing of the Magna Carta

    d) The Viking invasions

12. Who was Charlemagne?

    a) King of England

    b) Emperor of Byzantium

    c) King of the Franks who united much of Western Europe

d) The first Holy Roman Emperor

13. What was the main purpose of the Crusades?

    a) To establish trade routes

    b) To reclaim Jerusalem and other holy lands from Muslim control

    c) To spread Christianity to the Americas

    d) To overthrow the Byzantine Empire

14. What was the Magna Carta, and why was it significant?

    a) A Roman document that ended gladiatorial games

    b) An English charter that limited the powers of the king

    c) A French treaty that ended the Hundred Years' War

    d) A Spanish decree that expelled the Moors from Spain

15. Who was the founder of the Mongol Empire?

    a) Kublai Khan

    b) Genghis Khan

    c) Tamerlane

    d) Batu Khan

16. Who painted the Mona Lisa?

    a) Michelangelo

    b) Leonardo da Vinci

    c) Raphael

    d) Donatello

17. What invention by Johannes Gutenberg revolutionized the spread of information?

    a) The steam engine

    b) The printing press

    c) The telephone

    d) The telescope

18. Who wrote the 95 Theses, sparking the Protestant Reformation?

    a) John Calvin

    b) Martin Luther

    c) Ulrich Zwingli

    d) Ignatius of Loyola

19. What was the significance of the Treaty of Westphalia?

    a) It ended the Thirty Years' War

    b) It established the League of Nations

    c) It ended the War of Spanish Succession

    d) It led to the unification of Germany

20. Who was the Italian explorer credited with discovering the New World in 1492?

    a) Marco Polo

    b) Amerigo Vespucci

    c) Christopher Columbus

    d) Ferdinand Magellan

21. Who was the first European to reach India by sea?

    a) Vasco da Gama

    b) Ferdinand Magellan

    c) Christopher Columbus

    d) Hernán Cortés

22. What was the Columbian Exchange?

    a) A trade agreement between Spain and Portugal

    b) The transfer of plants, animals, and diseases between the Old World and the New World c) The establishment of the first European colonies in America

    d) The exploration of the Americas by Europeans

23. Which explorer's crew was the first to circumnavigate the globe?

    a) Christopher Columbus

    b) Vasco da Gama

    c) Ferdinand Magellan

    d) James Cook

24. Who was the conquistador responsible for the fall of the Aztec Empire?

    a) Francisco Pizarro

    b) Hernán Cortés

    c) Pedro Álvares Cabral

    d) Bartolomé de las Casas

25. What was the primary purpose of the Spanish Armada?

a) To establish trade routes with India

b) To overthrow Queen Elizabeth, I of England

c) To explore the New World

d) To conquer the Ottoman Empire

26. Who wrote "The Wealth of Nations"?

    a) John Locke

    b) Adam Smith

    c) Jean-Jacques Rousseau

    d) Voltaire

27. What was the main idea of John Locke's political philosophy?

    a) Absolute monarchy is the best form of government

    b) People have natural rights to life, liberty, and property

    c) The separation of powers is necessary for a balanced government

    d) The social contract must be enforced by a strong government

28. What event began the French Revolution?

    a) The Battle of Waterloo

    b) The Tennis Court Oath

    c) The Storming of the Bastille

    d) The Reign of Terror

29. Who led the Haitian Revolution?

a) Simon Bolivar

b) Toussaint L'Ouverture

c) Jean-Jacques Dessalines

d) Pedro I

30. What was the Industrial Revolution, and where did it begin?

    a) A period of rapid industrialization and innovation that began in Germany

    b) A period of rapid industrialization and innovation that began in the United States

    c) A period of rapid industrialization and innovation that began in Britain

    d) A period of rapid industrialization and innovation that began in France

31. Who was the emperor of France during the Napoleonic Wars?

    a) Louis XVI

    b) Napoleon Bonaparte

    c) Charles X

    d) Louis Philippe

32. What was the main cause of the American Civil War?

    a) The issue of states' rights

    b) The issue of tariffs

    c) The issue of slavery

d) The issue of territorial expansion

33. What was the purpose of the Congress of Vienna?

    a) To establish a unified German state

    b) To restore stability and balance of power in Europe after the Napoleonic Wars

    c) To create the League of Nations

    d) To negotiate peace terms after World War I

34. Who was the British monarch during the Victorian Era?

    a) Queen Elizabeth I

    b) Queen Victoria

    c) Queen Anne

    d) Queen Mary I

35. What was the Meiji Restoration?

    a) The period of modernization and industrialization in China

    b) The period of modernization and industrialization in Japan

    c) The period of Western colonization in Africa

    d) The period of British rule in India

36. What was the significance of the Berlin Conference of 1884-85?

    a) The division of Africa among European powers

    b) The unification of Germany

    c) The signing of the Treaty of Versailles

d) The formation of the League of Nations

37. What event triggered the start of World War I?

   a) The assassination of Archduke Franz Ferdinand of Austria

   b) The invasion of Poland by Germany

   c) The sinking of the Lusitania

   d) The signing of the Treaty of Versailles

38. Who was the leader of the Bolsheviks during the Russian Revolution?

   a) Joseph Stalin

   b) Leon Trotsky

   c) Vladimir Lenin

   d) Tsar Nicholas II

39. What was the Treaty of Versailles, and what did it accomplish?

   a) It ended World War I and imposed heavy reparations on Germany

   b) It ended World War II and led to the formation of the United Nations

   c) It established the League of Nations

   d) It created the European Union

40. Who was the dictator of Germany during World War II?

   a) Adolf Hitler

   b) Benito Mussolini

41. What event led the United States to enter World War II?

    a) The invasion of Poland

    b) The attack on Pearl Harbor

    c) The Battle of Britain

    d) The signing of the Munich Agreement

42. What was the significance of the D-Day invasion?

    a) It marked the beginning of the Allied invasion of Nazi-occupied Europe

    b) It was the final battle of World War II

    c) It led to the fall of Berlin

    d) It ended the Pacific War

43. What was the primary goal of the Marshall Plan?

    a) To provide economic aid to rebuild Western Europe after World War II

    b) To establish the United Nations

    c) To promote decolonization

    d) To create the European Union

44. Who was the leader of the Soviet Union during the Cuban Missile Crisis?

    a) Joseph Stalin

    b) Leonid Brezhnev

    c) Nikita Khrushchev

    d) Mikhail Gorbachev

45. What was the significance of the fall of the Berlin Wall?

    a) It marked the end of World War II

    b) It signalled the collapse of communism in Eastern Europe

    c) It led to the reunification of Germany

    d) Both b and c

46. Who was the first president of the People's Republic of China?

    a) Sun Yat-sen

    b) Chiang Kai-shek

    c) Mao Zedong

    d) Deng Xiaoping

47. What was the main outcome of the Vietnam War?

    a) The division of Vietnam into North and South

    b) The reunification of Vietnam under communist control

    c) The establishment of a democratic government in Vietnam

    d) The continuation of the war into Cambodia

48. What event marked the end of apartheid in South Africa?

    a) The election of Nelson Mandela as president

    b) The release of Nelson Mandela from prison

    c) The end of the Boer Wars

d) The establishment of the South African Republic

49. Who was the first female Prime Minister of the United Kingdom?

   a) Theresa May

   b) Margaret Thatcher

   c) Angela Merkel

   d) Indira Gandhi

50. What ancient wonder was located in the city of Alexandria, Egypt?

   a) The Great Pyramid

   b) The Colossus of Rhodes

   c) The Lighthouse of Alexandria

   d) The Hanging Gardens of Babylon

51. What was the Silk Road?

   a) A network of trade routes connecting the East and West

   b) An ancient Roman road system

   c) A series of maritime routes in the Indian Ocean

   d) A road built by the Mongols

52. Who was the first person to walk on the moon?

   a) Yuri Gagarin

   b) Buzz Aldrin

   c) John Glenn

d) Neil Armstrong

53. What was the main purpose of the United Nations? a) To promote international cooperation and peace

    b) To establish a global currency

    c) To colonize new territories

    d) To create a single world government

54. What was the significance of the Magna Carta?

    a) A charter that limited the powers of the English king and established certain legal rights

    b) A treaty that ended the Hundred Years' War

    c) An agreement that established the British Empire d) A document that declared American independence

55. Who was the first president of the United States?

    a) Thomas Jefferson

    b) John Adams

    c) George Washington

    d) Abraham Lincoln

56. Which Indian leader is known for his role in the independence movement?

    a) Jawaharlal Nehru

    b) Indira Gandhi

    c) Mahatma Gandhi

    d) Sardar Patel

57. Who was the longest-serving Prime Minister of the United Kingdom?

    a) Winston Churchill

    b) Margaret Thatcher

    c) David Lloyd George

    d) Tony Blair

58. Who was the first black president of South Africa?

    a) F.W. de Klerk

    b) Nelson Mandela

    c) Thabo Mbeki

    d) Jacob Zuma

59. Who was the first Chancellor of the Federal Republic of Germany?

    a) Adolf Hitler

    b) Angela Merkel

    c) Konrad Adenauer

    d) Willy Brandt

60. Who was the Chinese leader that initiated the Cultural Revolution?

    a) Sun Yat-sen

    b) Chiang Kai-shek

    c) Mao Zedong

    d) Deng Xiaoping

61. What was the Battle of Hastings?

    a) A battle in 1066 that resulted in Norman conquest of England

    b) A major battle of the Hundred Years' War

    c) A battle between the Romans and the Goths

    d) A naval battle between Spain and England

62. Who won the Battle of Waterloo?

    a) Napoleon Bonaparte

    b) The Duke of Wellington

    c) The Russian Empire

    d) The Ottoman Empire

63. What was the significance of the Battle of Gettysburg?

    a) It was the final battle of the American Civil War

    b) It was a major turning point in the American Civil War

    c) It led to the capture of Washington, D.C.

    d) It resulted in a treaty with Native American tribes

64. What was the primary cause of the Thirty Years' War? a) Territorial disputes

    b) Religious conflict

    c) Economic competition

    d) Dynastic succession issues

65. What was the result of the Spanish-American War?

    a) The independence of Mexico

b) The defeat of the Spanish Armada

c) The Treaty of Paris and U.S. acquisition of territories

d) The establishment of a Spanish colony in Florida

66. What was the significance of the Korean War?

    a) It was the first major conflict of the Cold War

    b) It led to the reunification of Korea

    c) It resulted in a peace treaty between North and South Korea

    d) It ended with the fall of Pyongyang

67. What was the Great Depression?

    a) A period of rapid economic growth in the 1920s

    b) A global economic downturn in the 1930s

    c) A period of inflation and high unemployment in the 1970s

    d) A financial crisis that began in 2008

68. What was the significance of the Bretton Woods Conference?

    a) It established the gold standard

    b) It created the World Bank and the International Monetary Fund (IMF)

    c) It ended the Great Depression

    d) It led to the formation of the European Union

69. What was the main cause of the 2008 financial crisis? a) The dot-com bubble

b) The housing market collapse and subprime mortgage crisis

c) The Asian financial crisis

d) The oil embargo

70. Who was the economist behind the theory of Keynesian economics?

    a) Milton Friedman

    b) Adam Smith

    c) John Maynard Keynes

    d) Friedrich Hayek

71. What was the primary purpose of the World Trade Organization (WTO)?

    a) To regulate international trade and resolve trade disputes

    b) To provide economic aid to developing countries

    c) To manage global financial markets

    d) To promote environmental sustainability

72. What was the significance of the Industrial Revolution?

    a) The transition from agrarian economies to industrialized economies

    b) The end of colonialism

    c) The establishment of the United Nations

    d) The discovery of the New World

73. What was the Harlem Renaissance?

a) A political movement in the 1920s

b) A cultural and artistic movement among African Americans in the 1920s

c) A scientific revolution in the 18th century

d) A religious revival in the 19th century

74. Who wrote "The Divine Comedy"?

    a) Geoffrey Chaucer

    b) William Shakespeare

    c) Dante Alighieri

    d) John Milton

75. What was the significance of the Renaissance?

    a) A period of religious reform and conflict

    b) A period of artistic, cultural, and intellectual revival in Europe

    c) A period of industrialization and urbanization

    d) A period of colonial expansion

76. Who was the composer of the Ninth Symphony?

    a) Johann Sebastian Bach

    b) Wolfgang Amadeus Mozart

    c) Ludwig van Beethoven

    d) Franz Schubert

77. What was the Enlightenment?

    a) A religious movement in the Middle Ages

b) A political movement advocating for democracy

c) An intellectual movement emphasizing reason and individualism

d) An artistic movement focused on romanticism

78. Who painted the Sistine Chapel ceiling?

    a) Raphael

    b) Leonardo da Vinci

    c) Donatello

    d) Michelangelo

79. Who developed the theory of relativity?

    a) Isaac Newton

    b) Albert Einstein

    c) Niels Bohr

    d) James Clerk Maxwell

80. What was the significance of the discovery of penicillin?

    a) It led to the development of antibiotics

    b) It revolutionized the field of genetics

    c) It was the first successful vaccine

    d) It paved the way for modern surgical techniques

81. Who was the first scientist to propose the heliocentric theory?

    a) Galileo Galilei

    b) Johannes Kepler

c) Nicolaus Copernicus

d) Isaac Newton

82. What was the Manhattan Project?

    a) A research project that developed the atomic bomb during World War II

    b) A space mission to explore the moon

    c) A construction project for the Manhattan skyline

    d) A global initiative to prevent nuclear proliferation

83. Who was the discoverer of the double helix structure of DNA?

    a) Charles Darwin

    b) James Watson and Francis Crick

    c) Gregor Mendel

    d) Louis Pasteur

84. What was the significance of the invention of the printing press?

    a) It made books more affordable and accessible

    b) It led to the Industrial Revolution

    c) It revolutionized transportation

    d) It was the first use of steam power

85. What was the Civil Rights Movement?

    a) A movement for workers' rights in the early 20th century

b) A movement in the 1950s and 1960s to end racial segregation and discrimination in the United States

c) A movement for women's suffrage in the 19th century

d) A movement to end colonial rule in Africa

86. Who led the Women's Suffrage Movement in the United States?

    a) Eleanor Roosevelt

    b) Sojourner Truth

    c) Elizabeth Cady Stanton and Susan B. Anthony

    d) Rosa Parks

87. What was the significance of the Stonewall Riots?

    a) They marked the beginning of the gay rights movement

    b) They ended segregation in schools

    c) They led to the passage of the Civil Rights Act

    d) They were a protest against the Vietnam War

88. What was the Arab Spring?

    a) A series of protests and uprisings in the Arab world that began in 2010

    b) A movement to establish a Jewish homeland in Palestine

    c) A cultural renaissance in the Middle East

    d) A campaign to modernize Arab economies

89. Who was the leader of the Indian independence movement?

   a) Jawaharlal Nehru

   b) Indira Gandhi

   c) Mahatma Gandhi

   d) Sardar Patel

90. What was the main goal of the Environmental Movement?

   a) To promote sustainable development and protect natural resources

   b) To end racial segregation

   c) To advance women's rights

   d) To improve working conditions for industrial labourers

91. What was the significance of the Cuban Missile Crisis?

   a) It brought the world to the brink of nuclear war

   b) It ended World War II

   c) It led to the fall of the Berlin Wall

   d) It marked the beginning of the Space Race

92. What was the primary cause of the Cold War?

   a) Territorial disputes between the United States and Soviet Union

   b) Ideological conflict between communism and capitalism

c) Competition for colonial territories

d) Disagreements over trade policies

93. What was the main outcome of the dissolution of the Soviet Union?

    a) The reunification of Germany

    b) The establishment of the European Union

    c) The end of the Cold War and the independence of multiple former Soviet states

    d) The beginning of World War III

94. What was the significance of the European Union's formation?

    a) It created a single European currency

    b) It established a common market for member states

    c) It promoted political and economic integration in Europe

    d) All of the above

95. What was the Treaty of Tordesillas?

    a) An agreement that divided the New World between Spain and Portugal

    b) A peace treaty that ended the Thirty Years' War

    c) A trade agreement between England and France

    d) An alliance between European powers against the Ottoman Empire

96. What was the Marshall Plan?

a) A U.S. program providing aid to Western Europe after World War II

b) A plan to build the Berlin Wall

c) A Soviet initiative to spread communism

d) A treaty to reduce nuclear arms

97. What was the significance of the fall of the Ottoman Empire?

    a) It led to the establishment of modern Turkey

    b) It marked the end of World War I

    c) It resulted in the creation of new nations in the Middle East

    d) All of the above

98. What was the primary purpose of the League of Nations?

    a) To promote international cooperation and prevent future wars

    b) To establish a global currency

    c) To colonize new territories

    d) To create a single world government

99. Who was the founder of modern Turkey?

    a) Mustafa Kemal Atatürk

    b) Suleiman the Magnificent

    c) Reza Shah

    d) Anwar Sadat

100. What was the significance of the Nuremberg Trials?
a) They brought Nazi war criminals to justice after World War II
b) They ended the Cold War
c) They established the European Union
d) They were the first trials for war crimes in history.

GEOGRAPHY QUIZ

1. Which is the largest continent by land area?

    a) Africa

    b) Asia

    c) Europe

    d) Antarctica

2. What is the capital of France?

    a) Berlin

    b) Rome

    c) Madrid

    d) Paris

3. Which ocean is the largest by surface area?

    a) Atlantic Ocean

    b) Indian Ocean

    c) Arctic Ocean

    d) Pacific Ocean

4. The Nile River flows into which sea?

    a) Mediterranean Sea

b) Red Sea

   c) Arabian Sea

   d) Black Sea

5. Which country has the largest population in the world?

   a) India

   b) China

   c) United States

   d) Indonesia

6. What is the smallest country in the world by land area?

   a) Monaco

   b) Nauru

   c) Vatican City

   d) San Marino

7. Mount Everest is located on the border of which two countries?

   a) India and China

   b) China and Nepal

   c) Nepal and Bhutan

   d) India and Bhutan

8. What is the capital of Japan?

   a) Beijing

   b) Tokyo

   c) Seoul

d) Bangkok

9. Which desert is the largest in the world?

   a) Sahara Desert

   b) Arabian Desert

   c) Gobi Desert

   d) Kalahari Desert

10. Which river is the longest in South America?

    a) Amazon River

    b) Orinoco River

    c) Paraná River

    d) São Francisco River

11. Which continent is known as the "Island Continent"?

    a) Asia

    b) Europe

    c) Australia

    d) Africa

12. The Great Barrier Reef is located off the coast of which country?

    a) Indonesia

    b) Australia

    c) Philippines

    d) Papua New Guinea

13. What is the capital of Canada?

a) Toronto

b) Vancouver

c) Ottawa

d) Montreal

14. The Andes Mountain range is located in which continent?

    a) Asia

    b) Europe

    c) Africa

    d) South America

15. The Amazon Rainforest is primarily located in which country?

    a) Brazil

    b) Peru

    c) Colombia

    d) Venezuela

16. What is the longest river in Africa?

    a) Congo River

    b) Nile River

    c) Zambezi River

    d) Niger River

17. Which country is known as the "Land of the Rising Sun"?

    a) China

b) Japan

c) South Korea

d) Thailand

18. Which U.S. state is the largest by land area?

    a) Texas

    b) California

    c) Alaska

    d) Montana

19. The city of Istanbul is located in which country?

    a) Greece

    b) Turkey

    c) Bulgaria

    d) Romania

20. Which European country is divided into cantons?

    a) Germany

    b) Switzerland

    c) Austria

    d) Belgium

21. What is the smallest ocean in the world?

    a) Atlantic Ocean

    b) Indian Ocean

    c) Southern Ocean

d) Arctic Ocean

22. The Galapagos Islands are part of which country?

    a) Peru

    b) Ecuador

    c) Chile

    d) Colombia

23. Which mountain range forms the natural border between France and Spain?

    a) Alps

    b) Pyrenees

    c) Carpathians

    d) Balkans

24. What is the currency used in Japan?

    a) Yen

    b) Won

    c) Yuan

    d) Ringgit

25. Which desert is located in southern Africa?

    a) Sahara Desert

    b) Gobi Desert

    c) Kalahari Desert

    d) Atacama Desert

26. Which country is both in Europe and Asia?

a) Russia

b) Turkey

c) Kazakhstan

d) All of the above

27. Which is the longest river in the United States?

    a) Mississippi River

    b) Missouri River

    c) Colorado River

    d) Ohio River

28. The Dead Sea is bordered by which two countries?
    a) Israel and Jordan

    b) Lebanon and Syria

    c) Egypt and Libya

    d) Turkey and Greece

29. What is the largest lake in Africa?

    a) Lake Tanganyika

    b) Lake Victoria

    c) Lake Malawi

    d) Lake Chad

30. Which country is known for its ancient pyramids?

    a) Greece

    b) Mexico

    c) Egypt

d) India

31. Which country has the most natural lakes?

    a) Canada

    b) United States

    c) Finland

    d) Russia

32. What is the deepest point in the world's oceans?

    a) Tonga Trench

    b) Mariana Trench

    c) Java Trench

    d) Puerto Rico Trench

33. The Ural Mountains are located in which country?

    a) China

    b) Kazakhstan

    c) Russia

    d) Mongolia

34. Which river flows through the Grand Canyon?

    a) Colorado River

    b) Rio Grande

    c) Mississippi River

    d) Columbia River

35. What is the highest waterfall in the world?

a) Niagara Falls

b) Victoria Falls

c) Angel Falls

d) Iguazu Falls

36. The island of Borneo is shared by how many countries?

    a) Two

    b) Three

    c) Four

    d) Five

37. Which desert covers much of Botswana?

    a) Sahara Desert

    b) Kalahari Desert

    c) Namib Desert

    d) Thar Desert

38. What is the capital of Bhutan?

    a) Thimphu

    b) Kathmandu

    c) Lhasa

    d) Dhaka

39. Which country is completely landlocked in South America?

    a) Bolivia

    b) Paraguay

c) Both a and b

d) Uruguay

40. The Azores are an autonomous region of which country?

    a) Spain

    b) Portugal

    c) Italy

    d) France

41. Which sea is the saltiest?

    a) Baltic Sea

    b) Red Sea

    c) Caspian Sea

    d) Mediterranean Sea

42. The highest peak in Africa, Mount Kilimanjaro, is located in which country?

    a) Kenya

    b) Tanzania

    c) Uganda

    d) Rwanda

43. Which European country has the most volcanoes?

    a) Iceland

    b) Italy

    c) Greece

d) Portugal

44. What is the longest river in Europe?

   a) Danube River

   b) Volga River

   c) Rhine River

   d) Thames River

45. The historic region of Mesopotamia is located in present-day:

   a) Iran

   b) Iraq

   c) Syria

   d) Turkey

46. Which country is known as the "Land of a Thousand Lakes"?

   a) Sweden

   b) Norway

   c) Finland

   d) Canada

47. The Ganges River is considered sacred in which religion?

   a) Islam

   b) Buddhism

   c) Hinduism

   d) Sikhism

48. What is the capital of Mongolia?

    a) Ulaanbaatar

    b) Astana

    c) Bishkek

    d) Dushanbe

49. Which U.S. state is known for its unique time zone, which is two hours behind Pacific Time?

    a) Alaska

    b) Hawaii

    c) Arizona

    d) Nevada

50. The Great Victoria Desert is located in which country?

    a) United States

    b) Argentina

    c) Australia

    d) Brazil

51. Which African lake is known for its high biodiversity and numerous endemic species?

    a) Lake Tanganyika

    b) Lake Victoria

    c) Lake Malawi

    d) Lake Chad

52. The city of Timbuktu, famous for its historical significance in trade and scholarship, is located in which country?

    a) Niger

    b) Mali

    c) Chad

    d) Mauritania

53. Which country has the longest coastline in the world?

    a) Australia

    b) Indonesia

    c) Canada

    d) Russia

54. The Strait of Hormuz connects which two bodies of water?

    a) The Red Sea and the Arabian Sea

    b) The Gulf of Oman and the Persian Gulf

    c) The Mediterranean Sea and the Black Sea

    d) The Atlantic Ocean and the Mediterranean Sea

55. The region of Transylvania is part of which country? a) Hungary

    b) Bulgaria

    c) Romania

    d) Serbia

56. Which mountain range is considered the oldest in the world?

    a) Rockies

    b) Urals

    c) Appalachians

    d) Himalayas

57. The Atacama Desert, one of the driest places on Earth, is located in which country?

    a) Argentina

    b) Chile

    c) Peru

    d) Bolivia

58. What is the capital of the country that is home to the ancient city of Petra?

    a) Amman

    b) Damascus

    c) Cairo

    d) Beirut

59. Which country is the largest producer of cocoa in the world?

    a) Ghana

    b) Indonesia

    c) Ivory Coast

    d) Brazil

60. The Sundarbans mangrove forest is located in which two countries?

    a) India and Bangladesh

    b) Myanmar and Thailand

    c) Indonesia and Malaysia

    d) Philippines and Vietnam

61. The Danakil Depression, known for its extreme heat and unique geological features, is located in which country?

    a) Sudan

    b) Ethiopia

    c) Kenya

    d) Chad

62. The historical city of Carthage was located in which present-day country?

    a) Libya

    b) Tunisia

    c) Algeria

    d) Morocco

63. Which river is the main source of water for Egypt?

    a) Nile River

    b) Tigris River

    c) Euphrates River

    d) Jordan River

64. Which country is known for having the world's most extensive system of inland waterways?

    a) United States

    b) Russia

    c) China

    d) Brazil

65. The Gulf Stream is an ocean current that primarily affects the climate of which continent?

    a) North America

    b) Europe

    c) South America

    d) Africa

66. Which country is known as the "Land of Smiles"?

    a) Thailand

    b) Indonesia

    c) Philippines

    d) Malaysia

67. The Bosporus Strait separates which two continents?

    a) Europe and Africa

    b) Asia and Africa

    c) Europe and Asia

    d) North America and South America

68. Which country has the highest number of UNESCO World Heritage Sites?

    a) China

    b) Italy

    c) Spain

    d) France

69. The Yucatán Peninsula is located in which country?

    a) Mexico

    b) Guatemala

    c) Belize

    d) Honduras

70. Which desert is known for its unique red sand dunes and is located in Namibia?

    a) Kalahari Desert

    b) Sahara Desert

    c) Namib Desert

    d) Gobi Desert

71. The historical region of Abyssinia is associated with which present-day country?

    a) Egypt

    b) Ethiopia

    c) Sudan

    d) Eritrea

72. Which river forms part of the border between the United States and Mexico?

    a) Mississippi River

    b) Colorado River

    c) Rio Grande

    d) Columbia River

73. The city of Reykjavik is the capital of which country?

    a) Norway

    b) Finland

    c) Iceland

    d) Greenland

74. Which African country was never colonized by European powers?

    a) Nigeria

    b) Ethiopia

    c) Ghana

    d) Kenya

75. Which country is home to the world's northernmost capital city?

    a) Norway

    b) Denmark

    c) Finland

    d) Iceland

76. The Taklamakan Desert is located in which country?

    a) Mongolia

    b) China

    c) Kazakhstan

    d) India

77. Which European city is known for its network of canals and is often referred to as the "Venice of the North"?

    a) Amsterdam

    b) Stockholm

    c) Bruges

    d) Copenhagen

78. What is the capital of the Canadian province of Quebec?

    a) Montreal

    b) Quebec City

    c) Ottawa

    d) Toronto

79. The Drakensberg Mountains are located in which country?

    a) South Africa

    b) Kenya

    c) Tanzania

    d) Namibia

80. The Mekong River flows through how many countries in Southeast Asia?

    a) Four

    b) Five

    c) Six

    d) Seven

81. The Rock of Gibraltar is a territory of which country?

    a) Spain

    b) Portugal

    c) United Kingdom

    d) France

82. Which desert is considered the coldest desert in the world?

    a) Gobi Desert

    b) Antarctic Desert

    c) Patagonian Desert

    d) Arctic Desert

83. Which country has the most islands in the world?

    a) Canada

    b) Indonesia

    c) Philippines

    d) Sweden

84. The ancient city of Tenochtitlan was located in which present-day country?

    a) Peru

    b) Mexico

    c) Colombia

    d) Guatemala

85. The Zambezi River is famous for which natural wonder?

    a) Victoria Falls

    b) Iguazu Falls

    c) Niagara Falls

    d) Angel Falls

86. The island of Socotra is part of which country?

    a) Somalia

    b) Yemen

    c) Oman

    d) Djibouti

87. Which river is the primary source of water for the city of Los Angeles?

    a) Sacramento River

    b) Colorado River

    c) San Joaquin River

    d) Rio Grande

88. Which country is known for the ancient rock-hewn churches of Lalibela?

    a) Sudan

    b) Egypt

    c) Ethiopia

    d) Libya

89. The Adriatic Sea is located between Italy and which other country?

    a) Greece

    b) Croatia

    c) Turkey

    d) Albania

90. Which U.S. state has the longest coastline? a) Florida

    b) California

    c) Alaska

    d) Texas

91. The ancient city of Troy is located in which present-day country?

    a) Greece

    b) Turkey

    c) Lebanon

    d) Egypt

92. The island of Tasmania is part of which country?

a) New Zealand

b) Australia

c) Indonesia

d) Papua New Guinea

93. The historical region of Bohemia is located in which present-day country?

    a) Poland

    b) Austria

    c) Czech Republic

    d) Germany

94. Which river flows through the city of Baghdad?

    a) Nile River

    b) Tigris River

    c) Euphrates River

    d) Jordan River

95. The ancient ruins of Machu Picchu are located in which mountain range?

    a) Andes

    b) Rockies

    c) Himalayas

    d) Alps

96. The city of Timbuktu is located near which river?

    a) Nile River

b) Niger River

c) Congo River

d) Senegal River

97. The historical region of Transylvania is located in which present-day country?

   a) Bulgaria

   b) Hungary

   c) Romania

   d) Serbia

98. The island of Corsica is part of which country?

   a) Italy

   b) Greece

   c) France

   d) Spain

99. The country of Lesotho is completely surrounded by which other country?

   a) Namibia

   b) Botswana

   c) South Africa

   d) Zimbabwe

100. The Amazon Rainforest spans across how many countries?

   a) Six

b) Eight

c) Nine

d) Ten

The Ultimate Quiz Book

MATHS QUIZ

21. Evaluate: 3+6×(5+4)÷3−7 .

    a) 11

    b) 12

    c) 13

    d) 14

22. Solve: (8−3)×2+6÷3.

    a) 8

    b) 9

    c) 10

    d) 11

23. Calculate: 5×(4+6)−10÷2 .

    a) 30

    b) 35

    c) 40

    d) 45

24. Evaluate: (9−4)÷(2+3)+8×2.

    a) 17

    b) 18

    c) 19

d) 20

25. Solve: 7+[8÷(4−2)×3].

    a) 15

    b) 16

    c) 17

    d) 18

26. Evaluate: 12÷(2×3)+4 .

    a) 6

    b) 7

    c) 8

    d) 9

27. Calculate: 10+(5×2)−8÷4 .

    a) 15

    b) 16

    c) 17

    d) 18

28. Evaluate: 25−[2×(3+2)]÷5 .

    a) 23

    b) 24

    c) 25

    d) 26

29. Solve: 18÷(2+1)×3 .

a) 18

b) 20

c) 22

d) 24

30. Calculate: (6+4)×(8−3)÷2.

    a) 20

    b) 25

    c) 30

    d) 35

31. Evaluate: 5+2×(3+4)−8÷2

    a) 10

    b) 11

    c) 12

    d) 13

32. Calculate: (6+3)×(2+5)÷7.

    a) 7

    b) 8

    c) 9

    d) 10

32. Solve: 20−[2×(5+3)÷4] .

    a) 14

    b) 15

c) 16

d) 17

33. Evaluate: 3×(8−5)+6÷2 .

    a) 12

    b) 13

    c) 14

    d) 15

34. Calculate: 9+(4×2−6÷3) .

    a) 13

    b) 14

    c) 15

    d) 16

35. Solve: 7+[9÷(3+6)×2].

    a) 7

    b) 8

    c) 9

    d) 10

36. Evaluate: (15÷3+4)×2.

    a) 10

    b) 12

    c) 14

    d) 16

37. Calculate: 24−(6×3)+8÷4.

   a) 7

   b) 8

   c) 9

   d) 10

38. Solve: 18+3×(4−2)÷6.

   a) 19

   b) 20

   c) 21

   d) 22

39. Evaluate: (8+2)×(5−3)÷4.

   a) 3

   b) 4

   c) 5

   d) 6

21. What is the simple interest on a principal of $2000 at a rate of 5% per year for 3 years?

   a) $200

   b) $250

   c) $300

   d) $350

22. If a sum of $1500 earns $225 as interest in 2 years, what is the rate of simple interest?

a) 7.5%

b) 8.0%

c) 8.5%

d) 9.0%

23. How long will it take for an amount of $900 to earn $180 as interest at a simple interest rate of 4% per year?

    a) 4 years

    b) 5 years

    c) 6 years

    d) 7 years

24. What is the principal amount if the simple interest earned in 5 years at 6% per annum is $450?

    a) $1200

    b) $1300

    c) $1400

    d) $1500

25. If $1200 is invested at a rate of 4% per year for 3 years, what is the total amount at the end of the period?

    a) $1320

    b) $1330

    c) $1340

    d) $1350

26. Calculate the simple interest on $5000 at a rate of 3% per year for 4 years.

a) $600

b) $700

c) $800

d) $900

27. A sum of money earns $1200 as interest in 3 years at a simple interest rate of 8% per annum. What is the principal amount?

    a) $4000

    b) $4500

    c) $5000

    d) $5500

28. What is the simple interest on a principal of $2500 at a rate of 6% per year for 5 years?

    a) $600

    b) $700

    c) $750

    d) $800

29. If $1800 earns $324 as interest in 3 years, what is the rate of simple interest?

    a) 5%

    b) 6%

    c) 7%

    d) 8%

30. How long will it take for an amount of $1600 to double at a simple interest rate of 5% per year?

    a) 10 years

    b) 15 years

    c) 20 years

    d) 25 years

31. What is the compound interest on $2000 at a rate of 4% per year, compounded annually, for 2 years?

    a) $160

    b) $163.20

    c) $164.80

    d) $168.20

32. If $3000 is compounded semi-annually at a rate of 6% per year, what will be the amount after 1 year?

    a) $3120.00

    b) $3121.80

    c) $3122.50

    d) $3123.60

33. What is the effective annual rate if the nominal rate is 5% compounded quarterly?

    a) 5.06%

    b) 5.09%

    c) 5.12%

    d) 5.15%

34. Calculate the compound interest on $2500 at a rate of 3% per year, compounded annually, for 3 years.

    a) $222.60

    b) $228.30

    c) $232.50

    d) $235.80

35. If $1800 is compounded monthly at a rate of 5% per year, what will be the amount after 2 years?

    a) $1980.20

    b) $1992.40

    c) $2005.30

    d) $2011.50

36. What is the compound interest on $1500 at a rate of 6% per year, compounded annually, for 4 years?

    a) $354.92

    b) $356.78

    c) $360.30

    d) $362.40

37. If $2200 is compounded quarterly at a rate of 4% per year, what will be the amount after 2 years?

    a) $2350.00

    b) $2354.60

    c) $2358.40

    d) $2362.50

38. What is the effective annual rate if the nominal rate is 7% compounded monthly?

    a) 7.11%

    b) 7.17%

    c) 7.19%

    d) 7.21%

39. Calculate the compound interest on $3500 at a rate of 8% per year, compounded annually, for 5 years.

    a) $1574.90

    b) $1598.60

    c) $1608.40

    d) $1620.50

40. If $5000 is compounded semi-annually at a rate of 3% per year, what will be the amount after 3 years?

    a) $5460.90

    b) $5474.30

    c) $5480.00

    d) $5493.60

41. A trader buys an article for $250 and sells it for $300. What is his profit percentage?

    a) 15%

    b) 20%

    c) 25%

    d) 30%

42. If an article is sold for $150 at a loss of 10%, what was its cost price?

    a) $160

    b) $165

    c) $166.67

    d) $170

43. A shopkeeper sells a product for $180, making a profit of 20%. What was the cost price of the product?

    a) $140

    b) $145

    c) $150

    d) $155

44. If a man sells a bicycle for $240 at a profit of 25%, what was the cost price of the bicycle?

    a) $180

    b) $185

    c) $190

    d) $195

45. A trader marks an article at $400 and sells it for $360 after allowing a discount. What is the discount percentage?

    a) 8%

    b) 9%

    c) 10%

d) 12%

46. Calculate the selling price if an item is bought for $320 and sold at a profit of 15%.

    a) $350

    b) $355

    c) $360

    d) $368

47. If a product is sold for $180 and a loss of 10% is incurred, what was the cost price?

    a) $190

    b) $200

    c) $210

    d) $220

48. A trader makes a profit of 25% by selling an article for $250. What was the cost price of the article?

    a) $180

    b) $190

    c) $200

    d) $210

49. Calculate the percentage profit if an article is bought for $200 and sold for $230.

    a) 10%

    b) 12%

    c) 14%

d) 15%

50. If an article is sold at a profit of 15% and the selling price is $345, what was the cost price?

    a) $290

    b) $300

    c) $305

    d) $310

51. What is the discount on an article marked at $500, if a discount of 10% is given?

    a) $40

    b) $45

    c) $50

    d) $55

52. Calculate the selling price of an article marked at $600 after a discount of 12%.

    a) $520

    b) $528

    c) $530

    d) $540

53. If an article is marked at $800 and sold at $720, what is the discount percentage?

    a) 8%

    b) 9%

    c) 10%

d) 12%

54. A shopkeeper offers a discount of 15% on an article marked at $400. What is the selling price?

a) $330

b) $340

c) $350

d) $360

55. If a product is marked at $1000 and sold for $850, what is the discount percentage?

a) 12%

b) 13%

c) 14%

d) 15%

56. Calculate the discount on an article marked at $1200 if a discount of 18% is given.

a) $200

b) $210

c) $216

d) $220

57. What is the selling price of an article marked at $1500 after a discount of 20%?

a) $1200

b) $1220

c) $1240

d) $1260

58. An article is marked at $750 and sold at $675. What is the discount percentage?

   a) 8%

   b) 9%

   c) 10%

   d) 11%

59. A shopkeeper offers a discount of 25% on an article marked at $300. What is the selling price?

   a) $200

   b) $210

   c) $220

   d) $225

60. Calculate the selling price if an item is bought for $400 and sold at a discount of 15%.

   a) $320

   b) $330

   c) $340

   d) $350

61. If 6 workers can complete a task in 12 days, how many days will it take for 9 workers to complete the same task?

   a) 6 days

   b) 8 days

c) 10 days

d) 12 days

62. A and B together can complete a job in 8 days. B alone can do it in 12 days. How long will A alone take to complete the job?

    a) 18 days

    b) 20 days

    c) 22 days

    d) 24 days

63. If a pipe can fill a tank in 10 hours, but there is a leak that can empty it in 6 hours, how long will it take to fill the tank if both the pipe and the leak are working simultaneously?

    a) 20 hours

    b) 25 hours

    c) 30 hours

    d) 35 hours

64. C can complete a task in 15 days, and D can complete the same task in 20 days. If they work together, how long will it take to complete the task?

    a) 8 days

    b) 9 days

    c) 10 days

    d) 12 days

65. If 4 workers can complete a task in 16 days, how many workers are needed to complete the same task in 8 days?

a) 6 workers

b) 7 workers

c) 8 workers

d) 9 workers

66. A and B together can complete a job in 5 days. A alone can do it in 7 days. How long will B alone take to complete the job?

a) 10 days

b) 12 days

c) 14 days

d) 16 days

67. If a pipe can fill a tank in 8 hours, but there is a leak that can empty it in 12 hours, how long will it take to fill the tank if both the pipe and the leak are working simultaneously?

a) 16 hours

b) 18 hours

c) 20 hours

d) 24 hours

68. C can complete a task in 10 days, and D can complete the same task in 15 days. If they work together, how long will it take to complete the task?

a) 6 days

b) 7 days

c) 8 days

d) 9 days

69. If 3 workers can complete a task in 9 days, how many days will it take for 6 workers to complete the same task?

    a) 3 days

    b) 4 days

    c) 5 days

    d) 6 days

70. A and B together can complete a job in 10 days. B alone can do it in 15 days. How long will A alone take to complete the job?

    a) 20 days

    b) 22 days

    c) 24 days

    d) 25 days

71. If a car travels 150 kilometres in 3 hours, what is its average speed?

    a) 40 km/h

    b) 50 km/h

    c) 60 km/h

    d) 70 km/h

72. A train travels at a speed of 80 km/h for 2 hours. How far does it travel?

a) 120 km

b) 140 km

c) 160 km

d) 180 km

73. If a cyclist covers a distance of 45 kilometres in 3 hours, what is his average speed?

a) 10 km/h

b) 12 km/h

c) 15 km/h

d) 18 km/h

74. A boat travels 30 kilometres upstream in 3 hours and the same distance downstream in 2 hours. What is the boat's average speed?

a) 10 km/h

b) 12 km/h

c) 15 km/h

d) 18 km/h

75. If a runner completes a marathon (42.195 kilometres) in 3 hours, what is his average speed?

a) 12.5 km/h

b) 13.5 km/h

c) 14.0 km/h

d) 14.5 km/h

76. A train travels 240 kilometres in 4 hours. What is its average speed?

    a) 50 km/h

    b) 55 km/h

    c) 60 km/h

    d) 65 km/h

77. If a car travels at a speed of 90 km/h for 2.5 hours, how far does it travel?

    a) 200 km

    b) 210 km

    c) 220 km

    d) 225 km

78. A cyclist travels 72 kilometres in 4.5 hours. What is his average speed?

    a) 12 km/h

    b) 14 km/h

    c) 16 km/h

    d) 18 km/h

79. If a boat covers 20 kilometres downstream in 2 hours and the same distance upstream in 4 hours, what is its average speed?

    a) 5 km/h

    b) 6 km/h

c) 7 km/h

d) 8 km/h

80. A plane travels 480 kilometres in 1.5 hours. What is its average speed?

    a) 300 km/h

    b) 320 km/h

    c) 330 km/h

    d) 340 km/h

81. If a car travels at a speed of 75 km/h for 3.2 hours, how far does it travel?

    a) 220 km

    b) 230 km

    c) 240 km

    d) 250 km

82. A runner completes a 10-kilometer race in 40 minutes. What is his average speed?

    a) 12 km/h

    b) 13.5 km/h

    c) 14 km/h

    d) 15 km/h

83. If a train travels 360 kilometres in 6 hours, what is its average speed?

    a) 50 km/h

    b) 55 km/h

c) 60 km/h

d) 65 km/h

84. A cyclist covers a distance of 60 kilometres in 3.5 hours. What is his average speed?

    a) 15 km/h

    b) 16.5 km/h

    c) 17 km/h

    d) 18 km/h

85. If a boat travels 40 kilometres upstream in 4 hours and the same distance downstream in 3 hours, what is its average speed?

    a) 9.5 km/h

    b) 10 km/h

    c) 10.5 km/h

    d) 11 km/h

86. A plane travels 720 kilometres in 2 hours. What is its average speed?

    a) 350 km/h

    b) 360 km/h

    c) 370 km/h

    d) 380 km/h

87. If a car travels at a speed of 100 km/h for 1.8 hours, how far does it travel?

    a) 160 km

b) 170 km

c) 180 km

d) 190 km

88. A runner completes a 5-kilometer race in 25 minutes. What is his average speed?

    a) 10 km/h

    b) 11 km/h

    c) 12 km/h

    d) 13 km/h

89. If a train travels 600 kilometres in 5 hours, what is its average speed?

    a) 110 km/h

    b) 115 km/h

    c) 120 km/h

    d) 125 km/h

90. A cyclist covers a distance of 100 kilometres in 6 hours. What is his average speed?

    a) 15 km/h

    b) 16 km/h

    c) 17 km/h

    d) 18 km/h

91. If an item is bought for $500 and sold at a profit of 20%, what is the selling price?

a) $580

b) $600

c) $620

d) $640

92. Calculate the simple interest on $7500 at a rate of 7% per year for 3 years.

    a) $1500

    b) $1575

    c) $1600

    d) $1650

93. What is the compound interest on $8000 at a rate of 6% per year, compounded annually, for 3 years?

    a) $1450.72

    b) $1465.64

    c) $1476.72

    d) $1492.34

94. If 8 workers can complete a task in 5 days, how many days will it take for 4 workers to complete the same task?

    a) 5 days

    b) 8 days

    c) 10 days

    d) 12 days

95. If a trader marks an article at $650 and sells it for $585 after allowing a discount, what is the discount percentage?

a) 8%

b) 9%

c) 10%

d) 12%

96. A shopkeeper offers a discount of 12% on an article marked at $750. What is the selling price?

a) $660

b) $665

c) $670

d) $675

97. If a car travels at a speed of 80 km/h for 3.5 hours, how far does it travel?

a) 270 km

b) 280 km

c) 290 km

d) 300 km

98. A cyclist covers a distance of 90 kilometres in 5.5 hours. What is his average speed?

a) 15 km/h

b) 16.5 km/h

c) 17.5 km/h

d) 18.5 km/h

99. Calculate the compound interest on $6000 at a rate of 5% per year, compounded annually, for 4 years.

    a) $1260.84

    b) $1275.50

    c) $1282.04

    d) $1290.12

100. If $1000 is invested at a rate of 3% per year for 3 years, what is the total amount including simple interest at the end of the period?

    a) $1090

    b) $1095

    c) $1100

    d) $1110

PHYSICS QUIZ

1. What is the speed of light in a vacuum?

    a) 300,000 km/s

    b) 400,000 km/s

    c) 500,000 km/s

    d) 600,000 km/s

2. Which of the following is the unit of force?

    a) Newton

    b) Joule

    c) Watt

    d) Pascal

3. What is the formula for calculating kinetic energy?

    a) KE=12mv2

    b) KE=mg

    c) KE=mv

    d) KE=12mv

4. Which law states that for every action, there is an equal and opposite reaction?

   a) Newton's First Law

   b) Newton's Second Law

   c) Newton's Third Law

   d) Hooke's Law

5. What is the unit of electrical resistance?

   a) Ohm

   b) Ampere

   c) Volt

   d) Farad

6. What is the value of gravitational acceleration on Earth?

   a) 8.8 m/s²

   b) 9.8 m/s²

   c) 10.8 m/s²

   d) 11.8 m/s²

7. Which of the following is a scalar quantity?

   a) Velocity

b) Displacement

c) Force

d) Temperature

8. What is the frequency of a wave with a period of 0.5 seconds?

   a) 1 Hz

   b) 2 Hz

   c) 3 Hz

   d) 4 Hz

9. Which of the following is an example of a renewable energy source?

   a) Coal

   b) Natural Gas

   c) Wind Energy

   d) Oil

10. What is the unit of power?

    a) Joule

    b) Watt

    c) Newton

d) Pascal

11. What is the atomic number of carbon?

   a) 6

   b) 8

   c) 12

   d) 14

12. Which of the following is an acid?

   a) NaOH

   b) HCl

   c) KCl

   d) H2O

13. What is the chemical symbol for silver?

   a) Au

   b) Si

   c) Ag

   d) S

14. What is the atomic number of hydrogen?

   a) 1

b) 2

c) 3

d) 4

15. What is the pH of a neutral solution?

    a) 5

    b) 6

    c) 7

    d) 8

16. What is the formula for calculating gravitational potential energy?

    a) PE=mg

    b) PE=12mv^2

    c) PE=mgh

    d) PE=12kx^2

17. Which law states that the pressure of a gas is inversely proportional to its volume?

    a) Boyle's Law

    b) Charles's Law

    c) Avogadro's Law

d) Dalton's Law

18. What is the formula for calculating force?

    a) F=ma

    b) F=mv

    c) F=mg

    d) $F=12mv^2$

19. What is the unit of measurement for electric charge? a) Coulomb

    b) Joule

    c) Watt

    d) Newton

20. What is the unit of inductance?

    a) Farad

    b) Henry

    c) Tesla

    d) Weber

21. What is the energy stored in a spring called?

    a) Kinetic energy

    b) Potential energy

c) Elastic potential energy

d) Gravitational potential energy

22. What is the phenomenon in which light bends around an obstacle?

   a) Reflection

   b) Refraction

   c) Diffraction

   d) Dispersion

23. Who is known for the theory of relativity?

   a) Isaac Newton

   b) Albert Einstein

   c) Niels Bohr

   d) James Clerk Maxwell

24. What is the unit of magnetic field strength?

   a) Tesla

   b) Weber

   c) Henry

   d) Farad

25. What is the relationship between voltage (V), current (I), and resistance (R) in Ohm's Law?

    a) V=IR

    b) V=I^{2}R

    c) V=IRV

    d) V=IR^2

26. Which subatomic particle has a positive charge?

    a) Electron

    b) Neutron

    c) Proton

    d) Photon

27. Which of the following is an example of a transverse wave?

    a) Sound wave

    b) Light wave

    c) Water wave

    d) Radio wave

28. What is the force that opposes the motion of two surfaces sliding past each other?

    a) Gravitational force

b) Normal force

c) Friction

d) Tension

29. What is the unit of luminous intensity?

    a) Candela

    b) Lumen

    c) Lux

    d) Watt

30. What is the value of the universal gravitational constant (G)?

    a) 6.67×10^{-11} N m²/kg²

    b) 8.31×10^{-11} N m²/kg²

    c) 9.81×10^{-11} N m²/kg²

    d) 1.60×10^{-11} N m²/kg²

31. What is the term for the rate of change of velocity?

    a) Speed

    b) Acceleration

    c) Force

    d) Momentum

32. Which device is used to measure electric current?

    a) Voltmeter

    b) Ammeter

    c) Galvanometer

    d) Thermometer

33. What is the unit of electric potential difference?

    a) Coulomb

    b) Ohm

    c) Volt

    d) Ampere

34. What is the equation for Einstein's mass-energy equivalence?

    a) $E=mc^2$

    b) $E=mv^2$

    c) $E=12mv^2$

    d) $E=mgh$

35. Which of the following is an example of a scalar quantity?

    a) Force

b) Velocity

c) Acceleration

d) Mass

36. What is the unit of work?

    a) Joule

    b) Newton

    c) Watt

    d) Pascal

37. What is the primary force responsible for the tides on Earth?

    a) Gravitational force of the Sun

    b) Gravitational force of the Moon

    c) Electromagnetic force

    d) Nuclear force

38. What is the phenomenon in which a material can return to its original shape after being deformed?

    a) Plasticity

    b) Elasticity

    c) Ductility

d) Malleability

39. Which of the following is the smallest unit of matter that retains the properties of an element?

   a) Atom

   b) Molecule

   c) Electron

   d) Proton

40. What is the process of splitting a heavy nucleus into two lighter nuclei called?

   a) Fusion

   b) Fission

   c) Radioactive decay

   d) Beta decay

41. What is the SI unit of temperature?

   a) Celsius

   b) Fahrenheit

   c) Kelvin

   d) Rankine

42. Which law states that the total energy of an isolated system remains constant?

a) Law of Conservation of Mass

b) Law of Conservation of Energy

c) Law of Conservation of Momentum

d) Law of Conservation of Charge

43. What is the unit of capacitance?

   a) Ohm

   b) Henry

   c) Farad

   d) Weber

44. What is the frequency of a sound wave that has a wavelength of 2 meters and a speed of 340 meters per second?

   a) 150 Hz

   b) 170 Hz

   c) 180 Hz

   d) 190 Hz

45. What is the unit of magnetic flux?

   a) Tesla

   b) Weber

c) Henry

d) Farad

46. What is the primary cause of Earth's magnetic field?

   a) Movement of tectonic plates

   b) Rotation of Earth on its axis

   c) Convection currents in the Earth's outer core

   d) Gravitational pull of the Moon

47. What is the name of the force that acts on an object moving in a circular path, directed towards the centre of the circle?

   a) Centripetal force

   b) Centrifugal force

   c) Gravitational force

   d) Electromagnetic force

48. What is the unit of thermal conductivity?

   a) Watt per meter per Kelvin (W/m·K)

   b) Joule per meter per Kelvin (J/m·K)

   c) Pascal per meter per Kelvin (Pa/m·K)

   d) Newton per meter per Kelvin (N/m·K)

49. What is the term for the force per unit area exerted on a surface? a) Pressure b) Force c) Work d) Energy
50. What is the phenomenon in which an electric current generates a magnetic field?

   a) Electromagnetic induction

   b) Photoelectric effect

   c) Magnetic flux

   d) Thermoelectric effect

51. What is the speed of sound in air at room temperature?

   a) 150 m/s

   b) 240 m/s

   c) 340 m/s

   d) 420 m/s

52. What is the SI unit of luminous flux?

   a) Lumen

   b) Candela

   c) Lux

   d) Watt

53. Which phenomenon explains the bending of light as it passes through different mediums?

a) Reflection

b) Refraction

c) Diffraction

d) Interference

54. What is the primary function of a transformer in an electrical circuit?

    a) To convert AC to DC

    b) To change the voltage level

    c) To increase current flow

    d) To store electrical energy

55. What is the force that holds protons and neutrons together in an atomic nucleus?

    a) Electromagnetic force

    b) Gravitational force

    c) Strong nuclear force

    d) Weak nuclear force

56. What is the unit of electric field strength?

    a) Volt per meter (V/m)

    b) Newton per Coulomb (N/C)

c) Joule per Coulomb (J/C)

d) Both a and b

57. What is the measure of the disorder or randomness in a system?

    a) Enthalpy

    b) Entropy

    c) Free energy

    d) Internal energy

58. Which principle states that pressure applied to a confined fluid is transmitted undiminished throughout the fluid?

    a) Archimedes' Principle

    b) Pascal's Principle

    c) Bernoulli's Principle

    d) Torricelli's Principle

59. What is the unit of angular velocity?

    a) Radians per second (rad/s)

    b) Degrees per second (deg/s)

    c) Revolutions per minute (rpm)

    d) Both a and b

60. What is the primary factor that determines the speed of a wave on a string?

    a) The tension in the string

    b) The length of the string

    c) The mass per unit length of the string

    d) Both a and c

61. What is the phenomenon of producing an electromotive force (EMF) by changing the magnetic field?

    a) Electromagnetic radiation

    b) Electromagnetic induction

    c) Electric conduction

    d) Magnetic flux

62. What is the angle of incidence when a light ray strikes a plane mirror at an angle of 30 degrees to the normal?

    a) 30 degrees

    b) 60 degrees

    c) 90 degrees

    d) 0 degrees

63. What is the unit of the magnetic field in the CGS system?

a) Tesla

b) Gauss

c) Weber

d) Henry

64. Which type of radiation has the highest energy?

   a) Infrared radiation

   b) Ultraviolet radiation

   c) X-rays

   d) Gamma rays

65. What is the term for the amount of energy required to raise the temperature of 1 gram of water by 1 degree Celsius?

   a) Joule

   b) Calorie

   c) Watt

   d) BTU

66. What is the law that relates the pressure and volume of a gas at constant temperature?

   a) Boyle's Law

   b) Charles's Law

c) Avogadro's Law

d) Gay-Lussac's Law

67. What is the primary cause of the greenhouse effect?

    a) Carbon dioxide

    b) Oxygen

    c) Nitrogen

    d) Hydrogen

68. What is the term for the change in frequency or wavelength of a wave in relation to an observer who is moving relative to the wave source?

    a) Doppler effect

    b) Photoelectric effect

    c) Zeeman effect

    d) Compton effect

69. What is the unit of inductance in the SI system?

    a) Farad

    b) Henry

    c) Tesla

    d) Weber

70. What is the primary component of the Sun's energy output?

   a) Visible light

   b) Ultraviolet radiation

   c) Infrared radiation

   d) Gamma radiation

71. What is the term for the energy required to remove an electron from an atom?

   a) Ionization energy

   b) Electron affinity

   c) Electronegativity

   d) Binding energy

72. What is the process of heat transfer through direct contact of particles?

   a) Conduction

   b) Convection

   c) Radiation

   d) Evaporation

73. What is the law that states that the voltage induced in a circuit is directly proportional to the rate of change of the magnetic flux?

a) Faraday's Law

b) Lenz's Law

c) Ohm's Law

d) Ampere's Law

74. Which scientist is known for the discovery of the electron?

    a) Isaac Newton

    b) Albert Einstein
        c) J.J. Thomson

    d) Niels Bohr

75. What is the name of the effect where a material emits electrons when exposed to light?

    a) Doppler effect

    b) Photoelectric effect

    c) Zeeman effect

    d) Compton effect

76. What is the term for the point on a standing wave with zero displacement?

    a) Antinode

    b) Node

c) Crest

d) Trough

77. What is the unit of viscosity?

    a) Pascal-second (Pa·s)

    b) Newton-second (N·s)

    c) Joule-second (J·s)

    d) Watt-second (W·s)

78. What is the principle that states that the buoyant force on an object is equal to the weight of the fluid displaced by the object?

    a) Archimedes' Principle

    b) Pascal's Principle

    c) Bernoulli's Principle

    d) Torricelli's Principle

79. What is the term for the time it takes for half of a radioactive substance to decay?

    a) Half-life

    b) Decay constant

    c) Mean life

    d) Disintegration time

80. What is the unit of electric dipole moment?

    a) Coulomb-meter (C·m)

    b) Newton-meter (N·m)

    c) Joule-meter (J·m)

    d) Tesla-meter (T·m)

81. What is the phenomenon where two waves superpose to form a resultant wave of greater, lower, or the same amplitude?

    a) Diffraction

    b) Refraction

    c) Interference

    d) Polarization

82. What is the primary component of the Earth's core?

    a) Iron

    b) Nickel

    c) Silicon

    d) Magnesium

83. Which type of lens causes light rays to diverge?

    a) Convex lens

b) Concave lens

c) Cylindrical lens

d) Spherical lens

84. What is the law that states that the current through a conductor between two points is directly proportional to the voltage across the two points?

a) Faraday's Law

b) Lenz's Law

c) Ohm's Law

d) Ampere's Law

85. What is the process of energy transfer by electromagnetic waves?

a) Conduction

b) Convection

c) Radiation

d) Evaporation

86. What is the measure of the amount of matter in an object?

a) Weight

b) Mass

c) Density

d) Volume

87. What is the term for the resistance of a fluid to flow?

   a) Viscosity

   b) Conductivity

   c) Elasticity

   d) Permeability

88. What is the SI unit of radioactive decay?

   a) Curie (Ci)

   b) Becquerel (Bq)

   c) Rutherford (Rd)

   d) Sievert (Sv)

89. What is the phenomenon in which the direction of a wave changes as it passes from one medium to another?

   a) Reflection

   b) Refraction

   c) Diffraction

   d) Interference

90. What is the primary purpose of a capacitor in an electric circuit?

   a) To store energy

   b) To generate current

   c) To convert AC to DC

   d) To regulate voltage

91. What is the principle behind the functioning of a hydraulic lift?

   a) Bernoulli's Principle

   b) Archimedes' Principle

   c) Pascal's Principle

   d) Torricelli's Principle

92. What is the term for the speed at which an object falls through the air when the force of gravity is balanced by air resistance?

   a) Terminal velocity

   b) Free fall

   c) Escape velocity

   d) Projectile motion

93. Which law states that the total momentum of an isolated system remains constant?

   a) Law of Conservation of Mass

   b) Law of Conservation of Energy

   c) Law of Conservation of Momentum

d) Law of Conservation of Charge

94. What is the function of a diode in an electrical circuit?

   a) To store electrical energy

   b) To convert AC to DC

   c) To resist current flow

   d) To allow current to flow in one direction

95. What is the term for the bending of waves around the corners of an obstacle or through an aperture?

   a) Diffraction

   b) Refraction

   c) Reflection

   d) Interference

96. What is the unit of angular momentum?

   a) Joule-second (J·s)

   b) Newton-meter (N·m)

   c) Kilogram-meter per second (kg·m/s)

   d) Kilogram-meter squared per second (kg·m²/s)

97. What is the primary cause of tidal forces on Earth?

   a) Gravitational pull of the Sun

   b) Gravitational pull of the Moon

   c) Earth's rotation

   d) Ocean currents

98. What is the relationship between pressure and volume for a given amount of gas at constant temperature?

   a) Directly proportional

   b) Inversely proportional

   c) Unrelated

   d) Exponentially proportional

99. What is the principle that states that the buoyant force on an object is equal to the weight of the fluid displaced by the object?

   a) Archimedes' Principle

   b) Pascal's Principle

   c) Bernoulli's Principle

   d) Torricelli's Principle

100.   What is the term for the distance between two consecutive points in phase on a wave?

   a) Wavelength

   b) Frequency

   c) Amplitude

   d) Period

BIOLOGY QUIZ

1. Which of the following is the basic unit of life?

    a) Atom

    b) Molecule

    c) Cell

    d) Organism

2. What is the process by which plants make their own food?

    a) Respiration

    b) Photosynthesis

    c) Transpiration

    d) Germination

3. What is the powerhouse of the cell?

    a) Nucleus

    b) Mitochondria

    c) Ribosome

    d) Endoplasmic Reticulum

4. What is the genetic material found in the nucleus of a cell?

    a) RNA

    b) DNA

c) Protein

d) Lipid

5. Which blood cells are responsible for transporting oxygen?

    a) Red blood cells

    b) White blood cells

    c) Platelets

    d) Plasma

6. Which of the following is a primary consumer in a food chain?

    a) Lion

    b) Grasshopper

    c) Eagle

    d) Mushroom

7. What is the main function of chlorophyll in plants?

    a) Absorb water

    b) Absorb sunlight

    c) Store nutrients

    d) Transport minerals

8. Which of the following is not a carbohydrate?

    a) Glucose

    b) Starch

    c) Cellulose

    d) Glycerol

9. What is the process by which cells divide to form two daughter cells?

   a) Meiosis

   b) Mitosis

   c) Fertilization

   d) Germination

10. What is the term for a group of similar cells that perform a specific function?

    a) Organ

    b) Tissue

    c) Organ system

    d) Organism

11. What is the name of the pigment that gives skin its colour?

    a) Haemoglobin

    b) Melanin

    c) Chlorophyll

    d) Carotene

12. Which organ is responsible for filtering blood in the human body?

    a) Liver

    b) Kidney

    c) Heart

    d) Stomach

13. What is the structural unit of the nervous system?

    a) Neuron

    b) Nephron

    c) Osteocyte

    d) Myocyte

14. What is the name of the process by which an organism takes in oxygen and releases carbon dioxide?

    a) Photosynthesis

    b) Respiration

    c) Transpiration

    d) Fermentation

15. Which hormone regulates blood sugar levels?

    a) Insulin

    b) Adrenaline

    c) Thyroxine

    d) Oestrogen

16. What is the term for the variety of life in a particular habitat or ecosystem?

    a) Biodiversity

    b) Biome

    c) Biosphere

    d) Biomass

17. Which of the following is a protein that speeds up chemical reactions in the body?

    a) Carbohydrate

    b) Lipid

    c) Enzyme

    d) Nucleic acid

18. What is the basic building block of proteins?

    a) Monosaccharides

    b) Fatty acids

    c) Amino acids

    d) Nucleotides

19. What is the name of the molecule that carries genetic information in living organisms?

    a) RNA

    b) DNA

    c) ATP

    d) NADPH

20. Which of the following is an example of a symbiotic relationship?

    a) Parasitism

    b) Predation

    c) Competition

    d) Mutualism

21. What is the term for an organism that can produce its own food using light or chemical energy?

    a) Heterotroph

    b) Autotroph

    c) Decomposer

    d) Consumer

22. Which part of the plant is responsible for absorbing water and nutrients from the soil?

    a) Stem

    b) Leaf

    c) Flower

    d) Root

23. What is the name of the process by which a caterpillar becomes a butterfly?

    a) Germination

    b) Metamorphosis

    c) Photosynthesis

    d) Respiration

24. Which of the following is a component of the human circulatory system?

    a) Brain

    b) Lungs

    c) Heart

    d) Stomach

25. What is the main function of white blood cells?

    a) Transport oxygen

    b) Fight infections

    c) Clot blood

    d) Store fat

26. Which of the following is a type of connective tissue?

    a) Muscle

    b) Bone

    c) Skin

    d) Hair

27. What is the primary function of the large intestine in the human body?

    a) Absorb nutrients

    b) Digest food

    c) Absorb water

    d) Store bile

28. Which part of the cell is responsible for controlling its activities?

    a) Cytoplasm

    b) Mitochondria

    c) Nucleus

    d) Ribosome

29. What is the term for the movement of water across a semi-permeable membrane?

    a) Diffusion

    b) Osmosis

    c) Active transport

    d) Filtration

30. Which of the following is a fat-soluble vitamin?

    a) Vitamin C

    b) Vitamin B12

    c) Vitamin D

    d) Vitamin B6

31. What is the term for an organism's observable characteristics?

    a) Genotype

    b) Phenotype

    c) Genomics

    d) Genome

32. Which organelle is responsible for protein synthesis?

    a) Golgi apparatus

    b) Lysosome

    c) Ribosome

    d) Chloroplast

33. What is the term for the study of insects?

a) Herpetology

b) Ornithology

c) Entomology

d) Ichthyology

34. What is the primary function of the lymphatic system?

    a) Transport oxygen

    b) Remove waste

    c) Fight infections

    d) Digest food

35. Which of the following is a characteristic of all living organisms?

    a) Ability to move

    b) Ability to reproduce

    c) Ability to see

    d) Ability to breathe

36. What is the name of the process by which plants lose water vapor through their leaves?

    a) Photosynthesis

    b) Respiration

    c) Transpiration

    d) Germination

37. Which organelle is known as the "garbage disposal" of the cell?

a) Mitochondria

b) Ribosome

c) Lysosome

d) Chloroplast

38. What is the term for the double-helix structure of DNA?

    a) Alpha helix

    b) Beta sheet

    c) Triple helix

    d) Double helix

39. Which of the following is a function of the skeletal system?

    a) Produce red blood cells

    b) Digest food

    c) Filter blood

    d) Store fat

40. What is the term for the sequence of changes in an ecosystem that regenerates or creates a community over time?

    a) Evolution

    b) Succession

    c) Mutation

    d) Natural selection

41. Which of the following is a macronutrient?

    a) Vitamin C

b) Iron

c) Protein

d) Calcium

42. What is the process by which green plants convert carbon dioxide and water into glucose and oxygen using sunlight?

    a) Respiration

    b) Photosynthesis

    c) Transpiration

    d) Digestion

43. Which of the following is a reproductive organ in plants?

    a) Leaf

    b) Root

    c) Flower

    d) Stem

44. What is the term for the bond formed between two amino acids?

    a) Hydrogen bond

    b) Ionic bond

    c) Peptide bond

    d) Covalent bond

45. Which of the following is an example of a decomposer?

    a) Grasshopper

    b) Snake

c) Mushroom

d) Rabbit

46. What is the primary function of the respiratory system?

    a) Transport nutrients

    b) Exchange gases

    c) Fight infections

    d) Produce hormones

47. Which of the following is a genetic disorder?

    a) Diabetes

    b) Hypertension

    c) Down syndrome

    d) Tuberculosis

48. What is the term for the smallest unit of an element that retains its chemical properties?

    a) Molecule

    b) Compound

    c) Atom

    d) Isotope

49. Which of the following is a function of the excretory system?

    a) Produce red blood cells

    b) Remove waste products

    c) Transport oxygen

    d) Store nutrients

50. What is the term for the study of the distribution and abundance of organisms and their interactions with the environment?

   a) Ecology

   b) Genetics

   c) Physiology

   d) Biochemistry

51. Which part of the brain controls balance and coordination?

   a) Cerebrum

   b) Cerebellum

   c) Brainstem

   d) Hypothalamus

52. What is the term for the symbiotic relationship where one organism benefits and the other is unaffected?

   a) Mutualism

   b) Parasitism

   c) Commensalism

   d) Predation

53. What is the primary pigment involved in photosynthesis?

   a) Chlorophyll

   b) Carotene

c) Xanthophyll

d) Anthocyanin

54. Which of the following is a characteristic of monocots?

   a) Two seed leaves

   b) Network of veins in leaves

   c) Vascular bundles in a ring

   d) Parallel veins in leaves

55. What is the main function of ribosomes?

   a) Lipid synthesis

   b) Protein synthesis

   c) Carbohydrate storage

   d) DNA replication

56. What is the basic structural and functional unit of the kidney?

   a) Alveolus

   b) Nephron

   c) Neuron

   d) Osteon

57. Which type of RNA carries the genetic code from DNA to the ribosome for protein synthesis?

   a) rRNA

   b) tRNA

   c) mRNA

   d) siRNA

58. Which part of the brain is responsible for regulating heart rate, breathing, and blood pressure?

   a) Cerebrum

   b) Cerebellum

   c) Brainstem

   d) Hypothalamus

59. What is the term for the involuntary response to a stimulus?

   a) Reflex

   b) Instinct

   c) Habit

   d) Learning

60. What is the role of haemoglobin in the blood?

   a) Fight infection

b) Clot blood

c) Transport oxygen

d) Store nutrients

61. Which of the following is a function of the human skeletal system?

    a) Produce hormones

    b) Store calcium

    c) Filter blood

    d) Produce enzymes

62. What is the term for the thread-like structures that carry genetic information in the nucleus of a cell?

    a) Genes

    b) Chromosomes

    c) Alleles

    d) Nucleotides

63. Which kingdom includes multicellular, photosynthetic organisms?

    a) Fungi

    b) Animalia

    c) Plantae

d) Protista

64. What is the name of the process by which cells engulf solid particles?

   a) Pinocytosis

   b) Phagocytosis

   c) Exocytosis

   d) Endocytosis

65. Which of the following is a prokaryotic organism?

   a) Fungi

   b) Bacteria

   c) Protozoa

   d) Algae

66. What is the primary component of the cell membrane?

   a) Proteins

   b) Nucleic acids

   c) Lipids

   d) Carbohydrates

67. Which hormone regulates the sleep-wake cycle in humans?

a) Insulin

b) Melatonin

c) Thyroxine

d) Cortisol

68. What is the function of the large intestine in the human body?

    a) Absorb nutrients

    b) Store bile

    c) Absorb water

    d) Produce enzymes

69. What is the term for a plant that has a life cycle of more than two years?

    a) Annual

    b) Biennial

    c) Perennial

    d) Seasonal

70. Which part of the human brain is responsible for higher cognitive functions such as thinking and planning?

    a) Cerebrum

b) Cerebellum

c) Brainstem

d) Hypothalamus

71. What is the main function of the respiratory system?

    a) Transport nutrients

    b) Exchange gases

    c) Fight infections

    d) Produce hormones

72. Which of the following is a function of the human liver?

    a) Produce bile

    b) Filter blood

    c) Store glycogen

    d) All of the above

73. What is the name of the process by which bacteria reproduce?

    a) Binary fission

    b) Budding

    c) Sporulation

d) Fragmentation

74. What is the function of the stomata in plant leaves?

    a) Absorb sunlight

    b) Exchange gases

    c) Store water

    d) Transport nutrients

75. Which of the following is a characteristic of fungi?

    a) Autotrophic

    b) Heterotrophic

    c) Prokaryotic

    d) Photosynthetic

76. What is the main function of the human circulatory system?

    a) Remove waste

    b) Transport nutrients and oxygen

    c) Produce hormones

    d) Store energy

77. Which organ in the human body is responsible for filtering waste from the blood?

a) Liver

b) Kidneys

c) Heart

d) Lungs

78. What is the term for a relationship where both organisms benefit?

    a) Parasitism

    b) Mutualism

    c) Commensalism

    d) Predation

79. Which part of the cell is responsible for producing ATP, the energy currency of the cell?

    a) Ribosome

    b) Mitochondria

    c) Chloroplast

    d) Golgi apparatus

80. What is the term for a characteristic that increases an organism's ability to survive and reproduce?

    a) Mutation

    b) Adaptation

c) Variation

d) Evolution

81. Which of the following is a function of the human skin?

    a) Protect against infection

    b) Regulate body temperature

    c) Produce vitamin D

    d) All of the above

82. What is the term for the study of heredity and the variation of inherited characteristics?

    a) Ecology

    b) Genetics

    c) Physiology

    d) Biochemistry

83. Which part of the human ear is responsible for detecting sound vibrations?

    a) Cochlea

    b) Eardrum

    c) Ossicles

d) semicircular canals

84. What is the term for the specialized tissue in plants that transports water and nutrients?

a) Epidermis

b) Cortex

c) Phloem

d) Xylem

85. What is the name of the pigment that gives plants their green colour?

a) Melanin

b) Carotene

c) Chlorophyll

d) Xanthophyll

86. Which of the following is a characteristic of vertebrates?

a) Exoskeleton

b) Notochord

c) Spinal cord

d) Radial symmetry

87. What is the function of the human immune system?

a) Transport oxygen

b) Fight infections

c) Produce hormones

d) Absorb nutrients

88. Which part of the flower is responsible for producing pollen?

    a) Stamen

    b) Pistil

    c) Sepal

    d) Petal

89. What is the name of the protein that helps blood clot?

    a) Haemoglobin

    b) Fibrinogen

    c) Albumin

    d) Globulin

90. Which of the following is a type of asexual reproduction?

    a) Meiosis

    b) Fertilization

c) Binary fission

d) Conjugation

91. What is the name of the process by which plants convert light energy into chemical energy?

   a) Photosynthesis

   b) Respiration

   c) Fermentation

   d) Chemosynthesis

92. Which of the following is a function of the human digestive system?

   a) Produce hormones

   b) Transport oxygen

   c) Break down food

   d) Fight infections

93. What is the primary function of the large intestine?

   a) Absorb nutrients

   b) Absorb water

   c) Store bile

   d) Produce enzymes

94. What is the term for the process by which an organism maintains a stable internal environment?

   a) Homeostasis

   b) Metabolism

   c) Adaptation

   d) Evolution

95. Which of the following is a function of the human skeleton?

   a) Protect vital organs

   b) Produce blood cells

   c) Store minerals

   d) All of the above

96. What is the term for the movement of water from an area of high concentration to an area of low concentration through a semi-permeable membrane?

   a) Diffusion

   b) Osmosis

   c) Active transport

   d) Filtration

97. Which organelle is responsible for packaging and transporting proteins within the cell?

a) Endoplasmic reticulum

b) Lysosome

c) Golgi apparatus

d) Nucleus

98. What is the primary function of red blood cells?

a) Fight infections

b) Clot blood

c) Transport oxygen

d) Store fat

99. Which of the following is a component of the human respiratory system?

a) Kidney

b) Liver

c) Lungs

d) Stomach

100. What is the term for the study of the structure and function of tissues?

a) Cytology

b) Histology

c) Anatomy

d) Gynaecology

SPORTS QUIZ

1. **Which country won the FIFA World Cup in 2018?**

    a) Brazil

    b) Germany

    c) France

    d) Argentina

2. **Who holds the record for the most Grand Slam titles in men's tennis?**

    a) Roger Federer

    b) Rafael Nadal

    c) Novak Djokovic

    d) Pete Sampras

3. **What sport is known as "America's pastime"?**

    a) Basketball

    b) Football

    c) Baseball

    d) Ice Hockey

4. **Which team won the NBA Championship in 2021?**

a) Los Angeles Lakers

b) Milwaukee Bucks

c) Brooklyn Nets

d) Phoenix Suns

5. **Which country has the most Olympic gold medals in swimming?**

    a) Australia

    b) Germany

    c) USA

    d) China

6. **Who won the ICC Cricket World Cup in 2019?**

    a) India

    b) Australia

    c) England

    d) New Zealand

7. **In which year was the first modern Olympic Games held?**

    a) 1896

    b) 1900

    c) 1912

    d) 1920

8. **Which sport uses the terms 'birdie' and 'eagle'?**

    a) Tennis

b) Golf

c) Badminton

d) Squash

9. **Who holds the record for the most goals in a single World Cup?**

    a) Pele

    b) Miroslav Klose

    c) Just Fontaine

    d) Ronaldo

10. **What is the maximum break in snooker?**

    a) 147

    b) 150

    c) 155

    d) 160

11. **In which sport is the Ryder Cup contested?**

    a) Tennis

    b) Golf

    c) Polo

    d) Sailing

12. **Who was the first woman to win six gold medals in a single Olympic Games?**

    a) Marion Jones

    b) Wilma Rudolph

c) Kristin Otto

d) Florence Griffith-Joyner

13. **What is the term for a score of three under par in golf?**

    a) Birdie

    b) Eagle

    c) Albatross

    d) Bogey

14. **Which football club has won the most European Cup/UEFA Champions League titles?**

    a) AC Milan

    b) Bayern Munich

    c) Liverpool

    d) Real Madrid

15. **Which country hosted the 2016 Summer Olympics?**

    a) China

    b) UK

    c) Brazil

    d) Russia

16. **What sport is Michael Phelps associated with?**

    a) Athletics

    b) Swimming

    c) Cycling

    d) Gymnastics

17. **Which cricketer scored 100 centuries in international cricket?**

    a) Ricky Ponting

    b) Brian Lara

    c) Sachin Tendulkar

    d) Jacques Kallis

18. **Who is known as the 'King of Clay' in tennis?**

    a) Roger Federer

    b) Andy Murray

    c) Rafael Nadal

    d) Novak Djokovic

19. **Which country has won the most Rugby World Cups?**

    a) England

    b) Australia

    c) New Zealand

    d) South Africa

20. **In which sport can you perform a 'slam dunk'?**

    a) Volleyball

    b) Basketball

    c) Handball

    d) Netball

21. **Which sport is known as the "king of sports"?**

    a) Basketball

b) Soccer (Football)

c) Tennis

d) Cricket

22. **In which year did India win its first Cricket World Cup?**

    a) 1975

    b) 1983

    c) 1987

    d) 1992

23. **Who is the all-time leading scorer in NBA history?**

    a) Michael Jordan

    b) Kareem Abdul-Jabbar

    c) LeBron James

    d) Kobe Bryant

24. **Which team won the first-ever Super Bowl?**

    a) Green Bay Packers

    b) New England Patriots

    c) Dallas Cowboys

    d) Pittsburgh Steelers

25. **Which tennis player has won the most French Open titles?**

    a) Novak Djokovic

    b) Bjorn Borg

    c) Rafael Nadal

d) Roger Federer

26. **What is the national sport of Japan?**

    a) Sumo wrestling

    b) Baseball

    c) Judo

    d) Karate

27. **Who is known as the "Fastest Man Alive"?**

    a) Usain Bolt

    b) Tyson Gay

    c) Yohan Blake

    d) Carl Lewis

28. **Which country has won the most Davis Cup titles in tennis?**

    a) Australia

    b) France

    c) USA

    d) Spain

29. **In which sport is the term 'home run' used?**

    a) Basketball

    b) Baseball

    c) Cricket

    d) Football

30. **Which country won the 2019 Rugby World Cup?**

a) New Zealand

b) England

c) South Africa

d) Australia

31. **In which sport do players use a 'puck'?**

    a) Lacrosse

    b) Hockey

    c) Polo

    d) Golf

32. **Who won the women's singles title at the Wimbledon Championships in 2021?**

    a) Serena Williams

    b) Naomi Osaka

    c) Simona Halep

    d) Ashleigh Barty

33. **Which country has won the most medals in the Winter Olympics?**

    a) Norway

    b) USA

    c) Germany

    d) Canada

34. **Who is the only athlete to win the Olympic marathon twice in consecutive years?**

a) Emil Zátopek

b) Abebe Bikila

c) Eliud Kipchoge

d) Haile Gebrselassie

35. **Which football team has won the most English Premier League titles?**

    a) Manchester City

    b) Liverpool

    c) Chelsea

    d) Manchester United

36. **In which year did Michael Jordan win his first NBA Championship?**

    a) 1989

    b) 1990

    c) 1991

    d) 1992

37. **What is the highest governing body of international football?**

    a) FIFA

    b) UEFA

    c) CONMEBOL

    d) AFC

38. **In which sport is the Stanley Cup awarded?**

a) Basketball

b) Ice Hockey

c) Rugby

d) Football

39. **Who was the first Indian to win an individual Olympic gold medal?**

    a) Abhinav Bindra

    b) Milkha Singh

    c) Sushil Kumar

    d) P. T. Usha

40. **Which sport features in the film "Field of Dreams"?**

    a) Football

    b) Baseball

    c) Basketball

    d) Golf

41. **Who is the captain of the Indian men's national cricket team as of 2024?**

    a) Virat Kohli

    b) Rohit Sharma

    c) KL Rahul

    d) Jasprit Bumrah

42. **Which country won the UEFA Euro 2020 tournament?**

    a) France

b) Italy

c) Spain

d) Germany

43. **Who won the men's singles title at the Australian Open in 2021?**

    a) Roger Federer

    b) Novak Djokovic

    c) Rafael Nadal

    d) Dominic Thiem

44. **Which sport is played on the largest pitch?**

    a) Football

    b) Cricket

    c) Rugby

    d) Baseball

45. **In which year did Tiger Woods win his first Masters Tournament?**

    a) 1997

    b) 1999

    c) 2001

    d) 2003

46. **Which country won the first-ever T20 Cricket World Cup?**

    a) India

b) Pakistan

   c) England

   d) Australia

47. **Which country has won the most FIFA Women's World Cup titles?**

   a) Germany

   b) USA

   c) Norway

   d) Japan

48. **Who was the first player to reach 100 centuries in first-class cricket?**

   a) WG Grace

   b) Don Bradman

   c) Jack Hobbs

   d) Brian Lara

49. **Which sport is associated with the term 'scrum'?**

   a) Rugby

   b) American Football

   c) Soccer

   d) Basketball

50. **In which city were the 2000 Summer Olympics held?**

   a) Athens

   b) Beijing

c) Sydney

d) London

ANSWERS

Answers - History

1. a) Hieroglyphics
2. c) Qin Shi Huang
3. b) Inca
4. c) A Babylonian law code
5. b) Sargon of Akkad
6. c) Socrates
7. c) Peloponnesian War
8. b) A small Greek force held off a much larger Persian army
9. b) Augustus Caesar
10. b) A period of peace and stability throughout the Roman Empire
11. a) The fall of the Western Roman Empire
12. c) King of the Franks who united much of Western Europe
13. b) To reclaim Jerusalem and other holy lands from Muslim control
14. b) An English charter that limited the powers of the king
15. b) Genghis Khan
16. b) Leonardo da Vinci

17. b) The printing press
18. b) Martin Luther
19. a) It ended the Thirty Years' War
20. c) Christopher Columbus
21. a) Vasco da Gama
22. b) The transfer of plants, animals, and diseases between the Old World and the New World
23. c) Ferdinand Magellan
24. b) Hernán Cortés
25. b) To overthrow Queen Elizabeth, I of England
26. b) Adam Smith
27. b) People have natural rights to life, liberty, and property
28. c) The Storming of the Bastille
29. b) Toussaint L'Ouverture
30. c) A period of rapid industrialization and innovation that began in Britain
31. b) Napoleon Bonaparte
32. c) The issue of slavery
33. b) To restore stability and balance of power in Europe after the Napoleonic Wars
34. b) Queen Victoria
35. b) The period of modernization and industrialization in Japan

36. a) The division of Africa among European powers
37. a) The assassination of Archduke Franz Ferdinand of Austria
38. c) Vladimir Lenin
39. a) It ended World War I and imposed heavy reparations on Germany
40. a) Adolf Hitler
41. b) The attack on Pearl Harbor
42. a) It marked the beginning of the Allied invasion of Nazi-occupied Europe
43. a) To provide economic aid to rebuild Western Europe after World War II
44. c) Nikita Khrushchev
45. d) Both b and c
46. c) Mao Zedong
47. b) The reunification of Vietnam under communist control
48. a) The election of Nelson Mandela as president
49. b) Margaret Thatcher
50. c) The Lighthouse of Alexandria
51. a) A network of trade routes connecting the East and West
52. d) Neil Armstrong
53. a) To promote international cooperation and peace

54. a) A charter that limited the powers of the English king and established certain legal rights

55. c) George Washington

56. c) Mahatma Gandhi

57. b) Margaret Thatcher

58. b) Nelson Mandela

59. c) Konrad Adenauer

60. c) Mao Zedong

61. a) A battle in 1066 that resulted in Norman conquest of England

62. b) The Duke of Wellington

63. b) A major turning point in the American Civil War

64. b) Religious conflict

65. c) The Treaty of Paris and U.S. acquisition of territories

66. a) It was the first major conflict of the Cold War

67. b) A global economic downturn in the 1930s

68. b) It created the World Bank and the International Monetary Fund (IMF)

69. b) The housing market collapse and subprime mortgage crisis

70. c) John Maynard Keynes

71. a) To regulate international trade and resolve trade disputes

72. a) The transition from agrarian economies to industrialized economies
73. b) A cultural and artistic movement among African Americans in the 1920s
74. c) Dante Alighieri
75. b) A period of artistic, cultural, and intellectual revival in Europe
76. c) Ludwig van Beethoven
77. c) An intellectual movement emphasizing reason and individualism
78. d) Michelangelo
79. b) Albert Einstein
80. a) It led to the development of antibiotics
81. c) Nicolaus Copernicus
82. a) A research project that developed the atomic bomb during World War II
83. b) James Watson and Francis Crick
84. a) It made books more affordable and accessible
85. b) A movement in the 1950s and 1960s to end racial segregation and discrimination in the United States
86. c) Elizabeth Cady Stanton and Susan B. Anthony
87. a) They marked the beginning of the gay rights movement
88. a) A series of protests and uprisings in the Arab world that began in 2010

89. c) Mahatma Gandhi

90. a) To promote sustainable development and protect natural resources

91. a) It brought the world to the brink of nuclear war

92. b) Ideological conflict between communism and capitalism

93. c) The end of the Cold War and the independence of multiple former Soviet states

94. d) All of the above

95. a) An agreement that divided the New World between Spain and Portugal

96. a) A U.S. program providing aid to Western Europe after World War II

97. d) All of the above

98. a) To promote international cooperation and prevent future wars

99. a) Mustafa Kemal Atatürk

100. a) They brought Nazi war criminals to justice after World War II

Answers – Geography

1. b) Asia
2. d) Paris
3. d) Pacific Ocean
4. a) Mediterranean Sea
5. b) China
6. c) Vatican City
7. b) China and Nepal
8. b) Tokyo
9. a) Sahara Desert
10. a) Amazon River
11. c) Australia
12. b) Australia
13. c) Ottawa
14. d) South America
15. a) Brazil
16. b) Nile River
17. b) Japan
18. c) Alaska
19. b) Turkey
20. b) Switzerland

21. d) Arctic Ocean

22. b) Ecuador

23. b) Pyrenees

24. a) Yen

25. c) Kalahari Desert

26. d) All of the above

27. b) Missouri River

28. a) Israel and Jordan

29. b) Lake Victoria

30. c) Egypt

31. a) Canada

32. b) Mariana Trench

33. c) Russia

34. a) Colorado River

35. c) Angel Falls

36. b) Three

37. b) Kalahari Desert

38. a) Thimphu

39. c) Both a and b

40. b) Portugal

41. b) Red Sea

42. b) Tanzania

43. a) Iceland

44. b) Volga River

45. b) Iraq

46. c) Finland

47. c) Hinduism

48. a) Ulaanbaatar

49. b) Hawaii

50. c) Australia

51. c) Lake Malawi

52. b) Mali

53. c) Canada

54. b) The Gulf of Oman and the Persian Gulf

55. c) Romania

56. c) Appalachians

57. b) Chile

58. a) Amman

59. c) Ivory Coast

60. a) India and Bangladesh

61. b) Ethiopia

62. b) Tunisia

63. a) Nile River

64. b) Russia

65. b) Europe

66. a) Thailand

67. c) Europe and Asia

68. b) Italy

69. a) Mexico

70. c) Namib Desert

71. b) Ethiopia

72. c) Rio Grande

73. c) Iceland

74. b) Ethiopia

75. d) Iceland

76. b) China

77. a) Amsterdam

78. b) Quebec City

79. a) South Africa

80. c) Six

81. c) United Kingdom

82. b) Antarctic Desert

83. d) Sweden

84. b) Mexico

85. a) Victoria Falls

86. b) Yemen

87. b) Colorado River

88. c) Ethiopia

89. b) Croatia

90. c) Alaska

91. b) Turkey

92. b) Australia

93. c) Czech Republic

94. b) Tigris River

95. a) Andes

96. b) Niger River

97. c) Romania

98. c) France

99. c) South Africa

100. c) Nine

Answers - Maths

1. c) 13
2. b) 9
3. b) 35
4. a) 17
5. d) 18
6. b) 7
7. c) 17
8. a) 23
9. a) 18
10. c) 30
11. c) 12
12. b) 8
13. c) 16
14. b) 13
15. d) 16
16. c) 9
17. b) 12
18. b) 8
19. a) 19
20. b) 4

21. c) $300

22. d) 9.0%

23. b) 5 years

24. d) $1500

25. d) $1350

26. c) $800

27. c) $5000

28. c) $750

29. b) 6%

30. c) 20 years

31. b) $163.20

32. d) $3123.60

33. d) 5.15%

34. c) $232.50

35. d) $2011.50

36. d) $362.40

37. d) $2362.50

38. c) 7.19%

39. b) $1598.60

40. b) $5474.30

41. b) 20%

42. c) $166.67

43. c) $150

44. a) $180

45. c) 10%

46. d) $368

47. b) $200

48. c) $200

49. d) 15%

50. b) $300

51. c) $50

52. b) $528

53. c) 10%

54. b) $340

55. d) 15%

56. c) $216

57. a) $1200

58. c) 10%

59. a) $200

60. a) $320

61. b) 8 days

62. d) 24 days

63. c) 30 hours

64. c) 10 days

65. c) 8
66. c) 8 workers
67. c) 14 days
68. a) 16 hours
69. b) 7 days
70. a) 3 days
71. c) 24 days
72. c) 60 km/h
73. c) 160 km
74. b) 15 km/h
75. b) 12 km/h
76. b) 14 km/h
77. c) 60 km/h
78. d) 225 km
79. c) 16 km/h
80. a) 5 km/h
81. a) 320 km/h
82. c) 240 km
83. c) 15 km/h
84. c) 60 km/h
85. c) 17 km/h
86. b) 10 km/h

87. b) 360 km/h

88. c) 180 km

89. c) 12 km/h

90. c) 120 km/h

91. b) 16 km/h

92. b) $600

93. b) $1575

94. c) $1476.72

95. c) 10 days

96. c) 10%

97. a) $660

98. b) 280 km

99. b) 16.5 km/h

100. a) $1260.84

101. b) $1095

Answers - Physics

1. a) 300,000 km/s
2. a) Newton
3. a) $KE = \frac{1}{2}mv^2$
4. c) Newton's Third Law
5. a) Ohm
6. b) 9.8 m/s^2
7. d) Temperature
8. b) 2 Hz
9. c) Wind Energy
10. b) Watt
11. a) 6
12. b) HCl
13. c) Ag
14. a) 1
15. c) 7
16. c) $PE = mgh$
17. a) Boyle's Law
18. a) F=ma
19. a) Coulomb
20. b) Henry

21. c) Elastic potential energy

22. c) Diffraction

23. b) Albert Einstein

24. a) Tesla

25. a) V=IR

26. c) Proton

27. b) Light wave

28. c) Friction

29. a) Candela

30. a) 6.67×10^{-11} N m²/kg²

31. b) Acceleration

32. b) Ammeter

33. c) Volt

34. a) $E=mc^2$

35. d) Mass

36. a) Joule

37. b) Gravitational force of the Moon

38. b) Elasticity

39. a) Atom

40. b) Fission

41. c) Kelvin

42. b) Law of Conservation of Energy

43. c) Farad

44. b) 170 Hz

45. b) Weber

46. c) Convection currents in the Earth's outer core

47. a) Centripetal force

48. a) Watt per meter per Kelvin (W/m·K)

49. a) Pressure

50. a) Electromagnetic induction

51. c) 340 m/s

52. a) Lumen

53. b) Refraction

54. b) To change the voltage level

55. c) Strong nuclear force

56. d) Both a and b

57. b) Entropy

58. b) Pascal's Principle

59. d) Both a and b

60. d) Both a and c

61. b) Electromagnetic induction

62. a) 30 degrees

63. b) Gauss

64. d) Gamma rays

65. b) Calorie

66. a) Boyle's Law

67. a) Carbon dioxide

68. a) Doppler effect

69. b) Henry

70. a) Visible light

71. a) Ionization energy

72. a) Conduction

73. a) Faraday's Law

74. c) J.J. Thomson

75. b) Photoelectric effect

76. b) Node

77. a) Pascal-second (Pa·s)

78. a) Archimedes' Principle

79. a) Half-life

80. a) Coulomb-meter (C·m)

81. c) Interference

82. a) Iron

83. b) Concave lens

84. c) Ohm's Law

85. c) Radiation

86. b) Mass

87. a) Viscosity

88. b) Becquerel (Bq)

89. b) Refraction

90. a) To store energy

91. c) Pascal's Principle

92. a) Terminal velocity

93. c) Law of Conservation of Momentum

94. d) To allow current to flow in one direction

95. a) Diffraction

96. d) Kilogram-meter squared per second (kg·m^2/s)

97. b) Gravitational pull of the Moon

98. b) Inversely proportional

99. a) Archimedes' Principle

100. a) Wavelength

Answers - Biology

1. c) Cell
2. b) Photosynthesis
3. b) Mitochondria
4. b) DNA
5. a) Red blood cells
6. b) Grasshopper
7. b) Absorb sunlight
8. d) Glycerol
9. b) Mitosis
10. b) Tissue
11. b) Melanin
12. b) Kidney
13. a) Neuron
14. b) Respiration
15. a) Insulin
16. a) Biodiversity
17. c) Enzyme
18. c) Amino acids
19. b) DNA
20. d) Mutualism
21. b) Autotroph
22. d) Root
23. b) Metamorphosis
24. c) Heart
25. b) Fight infections
26. b) Bone
27. c) Absorb water
28. c) Nucleus
29. b) Osmosis
30. c) Vitamin D

31. b) Phenotype
32. c) Ribosome
33. c) Entomology
34. c) Fight infections
35. b) Ability to reproduce
36. c) Transpiration
37. c) Lysosome
38. d) Double helix
39. a) Produce red blood cells
40. b) Succession
41. c) Protein
42. b) Photosynthesis
43. c) Flower
44. c) Peptide bond
45. c) Mushroom
46. b) Exchange gases
47. c) Down syndrome
48. c) Atom
49. b) Remove waste products
50. a) Ecology
51. b) Cerebellum
52. c) Commensalism
53. a) Chlorophyll
54. d) Parallel veins in leaves
55. b) Protein synthesis
56. b) Nephron
57. c) mRNA
58. c) Brainstem
59. a) Reflex
60. c) Transport oxygen
61. b) Store calcium
62. b) Chromosomes
63. c) Plantae
64. b) Phagocytosis

65. b) Bacteria
66. c) Lipids
67. b) Melatonin
68. c) Absorb water
69. c) Perennial
70. a) Cerebrum
71. b) Exchange gases
72. d) All of the above
73. a) Binary fission
74. b) Exchange gases
75. b) Heterotrophic
76. b) Transport nutrients and oxygen
77. b) Kidneys
78. b) Mutualism
79. b) Mitochondria
80. b) Adaptation
81. d) All of the above
82. b) Genetics
83. a) Cochlea
84. d) Xylem
85. c) Chlorophyll
86. c) Spinal cord
87. b) Fight infections
88. a) Stamen
89. b) Fibrinogen
90. c) Binary fission
91. a) Photosynthesis
92. c) Break down food
93. b) Absorb water
94. a) Homeostasis
95. d) All of the above
96. b) Osmosis
97. c) Golgi apparatus
98. c) Transport oxygen

99. c) Lungs
100. b) Histology

Answers - Sports

1. c) France
2. b) Rafael Nadal
3. c) Baseball
4. b) Milwaukee Bucks
5. c) USA
6. c) England
7. a) 1896
8. b) Golf
9. c) Just Fontaine
10. a) 147
11. b) Golf
12. c) Kristin Otto
13. c) Albatross
14. d) Real Madrid
15. c) Brazil
16. b) Swimming
17. c) Sachin Tendulkar
18. c) Rafael Nadal
19. c) New Zealand
20. b) Basketball
21. b) Soccer (Football)
22. b) 1983
23. b) Kareem Abdul-Jabbar
24. a) Green Bay Packers
25. c) Rafael Nadal
26. a) Sumo wrestling
27. a) Usain Bolt
28. c) USA
29. b) Baseball
30. c) South Africa

31. b) Hockey
32. d) Ashleigh Barty
33. a) Norway
34. b) Abebe Bikila
35. d) Manchester United
36. c) 1991
37. a) FIFA
38. b) Ice Hockey
39. a) Abhinav Bindra
40. b) Baseball
41. b) Rohit Sharma
42. b) Italy
43. b) Novak Djokovic
44. b) Cricket
45. a) 1997
46. a) India
47. b) USA
48. c) Jack Hobbs
49. a) Rugby
50. c) Sydney

REFERENCES

History

1. The History of Ancient Egypt by Bob Brier
2. The Cambridge Ancient History series
3. A History of Greece to 322 B.C. by N.G.L. Hammond
4. The Romans: From Village to Empire by Mary T. Boatwright et al.
5. A Short History of the Middle Ages by Barbara H. Rosenwein
6. The Penguin History of Medieval Europe by Maurice Keen
7. The Renaissance: A Short History by Paul Johnson
8. The Reformation: A History by Diarmaid MacCulloch
9. The Discoverers by Daniel J. Boorstin
10. Over the Edge of the World by Laurence Bergreen
11. The Enlightenment: A Very Short Introduction by John Robertson
12. The Age of Revolution: 1789-1848 by Eric Hobsbawm
13. The Penguin History of Europe by J.M. Roberts
14. Napoleon: A Life by Andrew Roberts
15. The First World War by John Keegan
16. The Second World War by Antony Beevor
17. The Cold War: A New History by John Lewis Gaddis

18. Modern Times: The World from the Twenties to the Nineties by Paul Johnson
19. A History of the Modern World by R.R. Palmer and Joel Colton
20. The Ascent of Money by Niall Ferguson

Mathematics

1. "Mathematics for Class 10" by R.D. Sharma
2. "Quantitative Aptitude for Competitive Examinations" by R.S. Aggarwal
3. "Business Mathematics" by Dr. Qazi Zameeruddin and Dr. Vijay K. Khanna
4. "Speed Mathematics: Secrets of Lightning Mental Calculation" by Bill Handley
5. "Financial Mathematics: A Comprehensive Treatment" by Giuseppe Campolieti and Roman N. Makarov

Science

1. "Physics Concept Questions and Answers", Byjus.
2. Physics NEET Practice Questions, MCQs, Past Year Questions.
3. 500+ Solved Physics Homework and Exam Problems
4. Biology - A-Z Common Reference Questions for Academic Librarians, Cambridge University Press.
5. "The Sports Quiz Book" by Derek O'Brien
6. "The Complete Indian Sports Quiz Book" by Vijayan Bala
7. "A Question of Sport Quiz Book" by Gareth Edwards

Sports

1. "The Sports Quiz Book" by Derek O'Brien

2. "The Complete Indian Sports Quiz Book" by Vijayan Bala
3. "The Greatest Sports Quiz" by Berty and Akhila
4. "Famous Sports Personalities Quiz And Answers MCQs"

GLOSSARY

1. **Artifact**: An object made by a human being, typically an item of cultural or historical interest.
2. **Dynasty**: A line of hereditary rulers of a country.
3. **Revolution**: A forcible overthrow of a government or social order, in favor of a new system.
4. **Colony**: A country or area under the full or partial political control of another country, occupied by settlers from that country.
5. **Renaissance**: A period in Europe from the 14th to the 17th century, marked by a revival of art, culture, and learning.
6. **Imperialism**: A policy of extending a country's power and influence through diplomacy or military force.
7. **Feudalism**: A medieval European political system in which a lord granted land to vassals in exchange for military service and loyalty.
8. **Magna Carta**: A charter of liberties agreed upon by King John of England in 1215, considered foundational in the development of constitutional law.
9. **Heresy**: Belief or opinion contrary to orthodox religious doctrine.
10. **Inquisition**: A group of institutions within the Catholic Church whose aim was to combat heresy.

11. **Treaty of Versailles**: The peace treaty signed in 1919 that ended World War I and imposed harsh penalties on Germany.

12. **Industrial Revolution**: The transition to new manufacturing processes in Europe and the US in the 18th and 19th centuries.

13. **Cold War**: The period of geopolitical tension between the Soviet Union and the United States after World War II.

14. **Apartheid**: A policy or system of segregation or discrimination on grounds of race in South Africa.

15. **Crusades**: A series of religious wars initiated, supported, and sometimes directed by the Latin Church in the medieval period.

16. **Manhattan Project**: The research and development project during World War II that produced the first nuclear weapons.

17. **Civil Rights Movement**: The struggle for social justice that took place mainly during the 1950s and 1960s among African Americans to gain equal rights under the law in the United States.

18. **Fascism**: A political philosophy, movement, or regime that exalts nation and often race above the individual, with a centralized autocratic government.

19. **Samurai**: The military nobility and officer caste of medieval and early-modern Japan.

20. **Ming Dynasty**: A major dynasty that ruled China from the mid-14th to the mid-17th century.

21. **Meiji Restoration**: The political revolution in Japan in 1868 that brought about the final demise of the Tokugawa shogunate.

22. **Renaissance Man**: A person with knowledge and skills in a number of different areas.

23. **Great Depression**: A severe worldwide economic depression that took place mostly during the 1930s.

24. **Holocaust**: The mass murder of six million Jews and others by the Nazis during World War II.

25. **Silk Road**: An ancient network of trade routes that connected the East and West.

26. **Vikings**: Norse explorers, warriors, merchants, and pirates who raided and settled in many parts of northwestern Europe in the late 8th to early 11th centuries.

27. **Sparta**: A prominent city-state in ancient Greece known for its military-oriented society.

28. **Heliocentric Theory**: The astronomical model in which the Earth and planets revolve around the Sun at the center of the Solar System.

29. **Czar**: The title of rulers or emperors of Russia before 1917.

30. **Pharaoh**: The common title of the monarchs of ancient Egypt.

31. **Propaganda**: Information, especially of a biased or misleading nature, used to promote a political cause or point of view.

32. **Renaissance Art**: Art produced during the Renaissance period characterized by a focus on realism, perspective, and classical themes.

33. **Mosaic**: A picture or pattern produced by arranging together small coloured pieces of hard material, such as stone, tile, or glass.

34. **Guild**: A medieval association of craftsmen or merchants, often having considerable power.

35. **Baroque**: A style of European architecture, music, and art of the 17th and 18th centuries that is characterized by ornate detail.

36. **Humanism**: A Renaissance cultural movement that turned away from medieval scholasticism and revived interest in ancient Greek and Roman thought.

37. **Suffrage**: The right to vote in political elections.

38. **Voyager**: An explorer or traveller.

39. **Sultan**: A Muslim sovereign.

40. **Caliphate**: An Islamic state under the leadership of an Islamic steward with the title of caliph.

41. **Yuan Dynasty**: The ruling dynasty of China established by Kublai Khan, leader of the Mongolian Borjigin clan.

42. **Medieval**: Relating to the Middle Ages.

43. **Palaeolithic**: The early phase of the Stone Age, lasting about 2.5 million years, when primitive stone implements were used.

44. **Neolithic**: The later part of the Stone Age, when ground or polished stone weapons and implements prevailed.

45. **Cartography**: The science or practice of drawing maps.
46. **Decolonization**: The action of changing from colonial to independent status.
47. **Nationalism**: Identification with one's own nation and support for its interests, especially to the exclusion or detriment of the interests of other nations.
48. **Caste System**: A class structure determined by birth, prevalent in India.
49. **Monarchy**: A form of government with a monarch at the head.
50. **Empire**: An extensive group of states or countries under a single supreme authority, formerly especially an emperor or empress.

Geography

1. **Archipelago**: A group of islands.
2. **Delta**: A landform at the mouth of a river where it meets an ocean, sea, lake, or reservoir, formed from the deposition of sediment carried by the river as the flow velocity decreases.
3. **Latitude**: The distance north or south of the equator, measured in degrees.
4. **Tundra**: A cold, treeless region where the subsoil is permanently frozen, found in the Arctic and on the tops of mountains.
5. **Volcano**: A rupture in the Earth's crust where molten lava, ash, and gases are ejected.
6. **Longitude**: The distance east or west of the prime meridian, measured in degrees.

7. **Glacier**: A slowly moving mass of ice formed by the accumulation and compaction of snow.
8. **Basin**: A natural depression on the earth's surface, typically containing water.
9. **Continent**: One of the main landmasses of the globe, usually reckoned as seven in number.
10. **Isthmus**: A narrow strip of land, bordered on both sides by water, connecting two larger bodies of land.
11. **Plateau**: An area of relatively level high ground.
12. **Equator**: An imaginary line around the middle of the Earth, equidistant from the North and South Poles.
13. **Peninsula**: A piece of land almost surrounded by water or projecting out into a body of water.
14. **Desert**: A barren area of landscape where little precipitation occurs and, consequently, living conditions are hostile for plant and animal life.
15. **Savanna**: A grassy plain in tropical and subtropical regions, with few trees.
16. **Rainforest**: A dense, tropical forest with a high amount of annual rainfall.
17. **Monsoon**: A seasonal wind in South and Southeast Asia, bringing heavy rains.
18. **Ocean**: A very large expanse of sea, one of the main divisions of the world's sea.
19. **River**: A large natural stream of water flowing in a channel to the sea, a lake, or another river.

20. **Mountain**: A large natural elevation of the earth's surface rising abruptly from the surrounding level.
21. **Canyon**: A deep gorge, typically one with a river flowing through it.
22. **Valley**: A low area of land between hills or mountains, typically with a river or stream flowing through it.
23. **Bay**: A broad inlet of the sea where the land curves inward.
24. **Gulf**: A deep inlet of the sea almost surrounded by land, with a narrow mouth.
25. **Strait**: A narrow passage of water connecting two seas or two large areas of water.
26. **Delta**: A landform at the mouth of a river where it meets an ocean, sea, lake, or reservoir, formed from the deposition of sediment carried by the river as the flow velocity decreases.
27. **Cape**: A headland of large size extending into a body of water, usually the sea.
28. **Atoll**: A ring-shaped reef, island, or chain of islands formed of coral.
29. **Fjord**: A long, narrow, deep inlet of the sea between high cliffs, typically formed by submergence of a glaciated valley.
30. **Oasis**: A fertile spot in a desert where water is found.
31. **Delta**: A landform at the mouth of a river formed from the deposition of sediment carried by the river.
32. **Globe**: A spherical representation of the Earth.

33. **Cartography**: The science or practice of drawing maps.

34. **Urbanization**: The process of making an area more urban.

35. **Erosion**: The process by which the surface of the earth gets worn down.

36. **Estuary**: The tidal mouth of a large river, where the tide meets the stream.

37. **Geology**: The science that deals with the Earth's physical structure and substance.

38. **Continent**: One of the Earth's main continuous expanses of land (Africa, Antarctica, Asia, Australia, Europe, North America, and South America).

39. **Hemisphere**: Half of the Earth, usually divided into northern and southern halves by the equator, or into western and eastern halves by an imaginary line passing through the poles.

40. **Biodiversity**: The variety of life in the world or in a particular habitat or ecosystem.

41. **Seismology**: The study of earthquakes and the propagation of elastic waves through the Earth.

42. **Savanna**: A grassy plain in tropical and subtropical regions, with few trees.

43. **Tropic of Cancer**: The parallel of latitude 23°26′ north of the equator.

44. **Tropic of Capricorn**: The parallel of latitude 23°26′ south of the equator.

45. **Longitude**: The angular distance of a place east or west of the meridian at Greenwich, England.

46. **Meridian**: A circle of constant longitude passing through a given place on the earth's surface and the terrestrial poles.

47. **Permafrost**: Ground that remains completely frozen for at least two years straight.

48. **Sediment**: Matter that settles to the bottom of a liquid; dregs.

49. **Subduction Zone**: A region of the Earth's crust where tectonic plates meet.

Physics

1. **Atom**: The basic unit of a chemical element, consisting of a nucleus of protons and neutrons with electrons orbiting around it.

2. **Quantum**: The minimum amount of any physical entity involved in an interaction, in quantum mechanics.

3. **Gravity**: The force that attracts a body towards the center of the Earth, or towards any other physical body having mass.

4. **Velocity**: The speed of something in a given direction.

5. **Energy**: The capacity to do work or produce change, in various forms such as kinetic, potential, thermal, electrical, chemical, nuclear, etc.

6. **Photon**: A particle representing a quantum of light or other electromagnetic radiation.

7. **Entropy**: A measure of the disorder or randomness in a system.

8. **Kinetics**: The branch of chemistry or biochemistry concerned with measuring and studying the rates of reactions.

9. **Magnetism**: The force exerted by magnets when they attract or repel each other.

10. **Nuclear Fusion**: A nuclear reaction in which atomic nuclei of low atomic number fuse to form a heavier nucleus with the release of energy.

11. **Nuclear Fission**: The splitting of a heavy atomic nucleus into two lighter nuclei, accompanied by the release of a large amount of energy.

12. **Thermodynamics**: The branch of physical science that deals with the relations between heat and other forms of energy.

13. **Oscillation**: Movement back and forth at a regular speed.

14. **Acceleration**: The rate of change of velocity of an object with respect to time.

15. **Inertia**: The resistance of any physical object to any change in its velocity.

16. **Refraction**: The change in direction of a wave passing from one medium to another caused by its change in speed.

17. **Diffraction**: The bending of waves around obstacles and openings.

18. **Resonance**: The reinforcement or prolongation of sound by reflection from a surface or by the synchronous vibration of a neighbouring object.

19. **Superconductivity**: A phenomenon of exactly zero electrical resistance and expulsion of magnetic fields occurring in certain materials when cooled below a characteristic critical temperature.

20. **Dark Matter**: A form of matter thought to account for approximately 85% of the matter in the universe, its presence inferred from its gravitational effects on visible matter.

21. **Relativity**: The dependence of various physical phenomena on relative motion of the observer and the observed objects, especially regarding the nature and behaviour of light, space, time, and gravity.

22. **Electromagnetism**: The interaction of electric currents or fields and magnetic fields.

23. **Coulomb**: The unit of electric charge in the International System of Units (SI).

24. **Circuit**: A complete and closed path around which a circulating electric current can flow.

25. **Ampere**: The unit of electric current in the International System of Units (SI).

26. **Watt**: The SI unit of power, equivalent to one joule per second.

27. **Pascal**: The SI unit of pressure, equal to one newton per square meter.

28. **Luminosity**: The intrinsic brightness of a celestial object.

29. **Polarization**: The orientation of the oscillations in the wave, usually with respect to an angle.

30. **Wave-Particle Duality**: The concept that every particle or quantum entity may be described as either a particle or a wave.

31. **Planck's Constant**: The constant of proportionality relating the energy of a photon to the frequency of its associated electromagnetic wave.

32. **Heisenberg Uncertainty Principle**: A fundamental theory in quantum mechanics that defines why a simultaneous measurement of a particle's position and momentum cannot be precisely determined.

33. **Schrodinger's Cat**: A thought experiment that presents a cat that may be simultaneously both alive and dead, depending on an earlier random event.

34. **Centripetal Force**: A force that acts on a body moving in a circular path and is directed towards the centre around which the body is moving.

35. **Thermal Conduction**: The transfer of heat energy through a material.

36. **Isotopes**: Variants of a particular chemical element that differ in neutron number, and consequently in nucleon number.

37. **Photonics**: The science and technology of generating, controlling, and detecting photons, particularly in the visible and near-infrared light spectrum.

38. **Quasar**: An extremely luminous active galactic nucleus, with its energy output equivalent to that of a trillion suns.

39. **Black Hole**: A region of space having a gravitational field so intense that no matter or radiation can escape.

40. **Supernova**: A powerful and luminous explosion of a star, often resulting in a neutron star or black hole.
41. **Neutrino**: A subatomic particle with a mass close to zero and half-integer spin, rarely interacting with normal matter.
42. **Redshift**: The phenomenon where light from an object is increased in wavelength, or shifted to the red end of the spectrum, indicative of the Doppler effect, particularly with celestial objects.
43. **Event Horizon**: A boundary in spacetime beyond which events cannot affect an outside observer, associated with black holes.
44. **Cosmic Microwave Background**: Radiation left over from an early stage of the universe in Big Bang cosmology.
45. **Higgs Boson**: An elementary particle in the Standard Model of particle physics, responsible for giving other particles their mass.
46. **Singularity**: A point in spacetime where density becomes infinite, such as at the centre of a black hole.
47. **Entropy**: A measure of the randomness or disorder in a system.
48. **Electrostatic Force**: The force between charged objects.
49. **Gravitational Constant**: The empirical physical constant involved in the calculation of gravitational effects.
50. **Kepler's Laws**: Three scientific laws describing the motion of planets around the Sun.

Biology

1. **Cell**: The smallest structural and functional unit of an organism, typically microscopic and consisting of cytoplasm and a nucleus enclosed in a membrane.

2. **DNA (Deoxyribonucleic Acid)**: The molecule that carries the genetic instructions for life, consisting of two long chains of nucleotides twisted into a double helix.

3. **Ecosystem**: A biological community of interacting organisms and their physical environment.

4. **Photosynthesis**: The process by which green plants and some other organisms use sunlight to synthesize foods with the help of chlorophyll from carbon dioxide and water.

5. **Mutation**: A change in the DNA sequence that may lead to a variation in the genetic code.

6. **Genotype**: The genetic constitution of an individual organism.

7. **Phenotype**: The set of observable characteristics of an individual resulting from the interaction of its genotype with the environment.

8. **Homeostasis**: The tendency of a system, especially the physiological system of higher animals, to maintain internal stability.

9. **Osmosis**: The process by which molecules of a solvent pass through a semipermeable membrane from a less concentrated solution into a more concentrated one.

10. **Enzyme**: A substance produced by a living organism that acts as a catalyst to bring about a specific biochemical reaction.

11. **Chromosome**: A thread-like structure of nucleic acids and protein found in the nucleus of most living cells, carrying genetic information in the form of genes.

12. **Allele**: One of two or more alternative forms of a gene that arise by mutation and are found at the same place on a chromosome.

13. **Mitosis**: A type of cell division that results in two daughter cells each having the same number and kind of chromosomes as the parent nucleus.

14. **Meiosis**: A type of cell division that results in four daughter cells each with half the number of chromosomes of the parent cell, as in the production of gametes and plant spores.

15. **Gene**: A unit of heredity that is transferred from a parent to offspring and is held to determine some characteristic of the offspring.

16. **Genome**: The complete set of genes or genetic material present in a cell or organism.

17. **Apoptosis**: The death of cells that occurs as a normal and controlled part of an organism's growth or development.

18. **Symbiosis**: Interaction between two different organisms living in close physical association, typically to the advantage of both.

19. **Microorganism**: A microscopic organism, especially a bacterium, virus, or fungus.

20. **Antibody**: A blood protein produced in response to and counteracting a specific antigen.

21. **Antigen**: A toxin or other foreign substance that induces an immune response in the body, especially the production of antibodies.

22. **Pathogen**: A bacterium, virus, or other microorganism that can cause disease.

23. **Metabolism**: The chemical processes that occur within a living organism in order to maintain life.

24. **Homeostasis**: The tendency of the body to seek and maintain a condition of balance or equilibrium within its internal environment, even when faced with external changes.

25. **Photosynthesis**: The process by which green plants and some other organisms use sunlight to synthesize foods with the help of chlorophyll from carbon dioxide and water.

26. **Cellular Respiration**: The set of metabolic reactions and processes that take place in the cells of organisms to convert biochemical energy from nutrients into adenosine triphosphate (ATP).

27. **Ecology**: The branch of biology that deals with the relations of organisms to one another and to their physical surroundings.

28. **Evolution**: The process by which different kinds of living organisms are thought to have developed and diversified from earlier forms during the history of the earth.

29. **Genetics**: The study of heredity and the variation of inherited characteristics.

30. **Hormone**: A regulatory substance produced in an organism and transported in tissue fluids such as blood or sap to stimulate specific cells or tissues into action.

31. **Enzyme**: A substance produced by a living organism which acts as a catalyst to bring about a specific biochemical reaction.

32. **Molecule**: A group of atoms bonded together, representing the smallest fundamental unit of a chemical compound that can take part in a chemical reaction.

33. **Plasmid**: A genetic structure in a cell that can replicate independently of the chromosomes, typically a small circular DNA strand in the cytoplasm of a bacterium.

34. **Ribosome**: A complex molecule made of ribosomal RNA molecules and proteins that form a factory for protein synthesis in cells.

35. **Neuron**: A specialized cell transmitting nerve impulses; a nerve cell.

36. **Tissue**: Any of the distinct types of material of which animals or plants are made, consisting of specialized cells and their products.

37. **Organ**: A part of an organism that is typically self-contained and has a specific vital function, such as the heart or liver in humans.

38. **Organ System**: A group of organs that work together to perform one or more functions in the body.

39. **Stem Cell**: An undifferentiated cell of a multicellular organism that is capable of giving rise to indefinitely

more cells of the same type, and from which certain other kinds of cell arise by differentiation.

40. **Antibiotic**: A medicine that inhibits the growth of or destroys microorganisms.

41. **Virus**: A small infectious agent that replicates only inside the living cells of an organism.

42. **Bacteria**: A member of a large group of unicellular microorganisms that have cell walls but lack organelles and an organized nucleus, including some that can cause disease.

43. **Protist**: Any of a diverse taxonomic group and especially a kingdom (Protista) of eukaryotic organisms that are unicellular and sometimes colonial or less often multicellular and that typically include the protozoans, most algae, and often some fungi (such as slime molds).

44. **Fungi**: Any of a group of spore-producing organisms feeding on organic matter, including molds, yeast, mushrooms, and toadstools.

45. **Photosynthesis**: The process by which green plants and some other organisms use sunlight to synthesize foods from carbon dioxide and water.

46. **Chlorophyll**: A green pigment, present in all green plants and in cyanobacteria, responsible for the absorption of light to provide energy for photosynthesis.

47. **Genome**: The complete set of genes or genetic material present in a cell or organism.

48. **Ecosystem**: A biological community of interacting organisms and their physical environment.

49. **Mutation**: A change in the DNA sequence that affects genetic information.
50. **Natural Selection**: The process whereby organisms better adapted to their environment tend to survive and produce more offspring.

Sports

1. **Albatross (Golf)**: A score of three strokes under par on a single hole.
2. **Birdie (Golf)**: A score of one stroke under par on a single hole.
3. **Grand Slam (Tennis)**: Winning all four major championships (Australian Open, French Open, Wimbledon, and US Open) in a single calendar year.
4. **Hat-Trick (Cricket/Football/Hockey)**: Achieving three successes of the same kind, such as three wickets by a bowler in cricket, or three goals by a player in football or hockey, in one game.
5. **Home Run (Baseball)**: A hit that allows the batter to make a complete circuit of the bases and score a run without being put out.
6. **Knockout (Boxing)**: A situation in which a boxer is unable to continue fighting because they have been knocked to the ground by their opponent.
7. **Penalty Kick (Football)**: A free kick at the goal from a point within the penalty area, awarded after a foul has been committed.
8. **Ryder Cup (Golf)**: A biennial men's golf competition between teams from Europe and the United States.

9. **Scrum (Rugby)**: A method of restarting play in rugby, involving players packing closely together with their heads down and attempting to gain possession of the ball.

10. **Slam Dunk (Basketball)**: A high jump shot in which the ball is thrust down through the hoop.

11. **Tiebreak (Tennis)**: A game played to decide the winner of a set when the score is tied at 6-6.

12. **Touchdown (American Football)**: A score in American football worth six points, achieved by carrying the ball across the opponent's goal line or catching it in the end zone.

13. **Yellow Card (Football)**: A caution issued by the referee to a player who has committed a foul or misconduct, signified by a yellow card.

14. **Zonal Marking (Football)**: A defensive strategy where each player is responsible for covering an area of the field rather than a specific opponent.

15. **Knockout (Boxing)**: A victory in which one fighter is knocked to the ground and unable to continue.

16. **Over (Cricket)**: A set of six consecutive deliveries bowled by the same bowler.

17. **Six (Cricket)**: A hit that sends the ball over the boundary without touching the ground, scoring six runs.

18. **Ace (Tennis)**: A serve that the opponent fails to touch, winning the point outright.

19. **Break Point (Tennis)**: A point which, if won by the receiver, causes them to win the game.

20. **Bunker (Golf)**: A hazard on a golf course consisting of a depression filled with sand.
21. **Carom (Billiards)**: A shot in billiards in which the cue ball strikes two object balls.
22. **Chicane (Motorsport)**: A sequence of tight alternating turns in opposite directions.
23. **Cyclo-Cross (Cycling)**: A form of bicycle racing involving off-road courses and obstacles.
24. **Decathlon (Athletics)**: An athletic event taking place over two days, in which each competitor takes part in the same prescribed ten events.
25. **Epee (Fencing)**: A type of sword, the largest and heaviest of the three weapons used in fencing.
26. **Grand Prix (Motorsport)**: One of a series of international racing competitions.
27. **Heptathlon (Athletics)**: An athletic contest consisting of seven events.
28. **Hurdles (Athletics)**: A race in which runners jump over a series of barriers, known as hurdles.
29. **Ironman (Triathlon)**: A long-distance triathlon race consisting of a 2.4-mile swim, a 112-mile bicycle ride, and a marathon 26.2-mile run

Notes

Unleash Your Inner Quiz Master with The Ultimate Quiz Book!

Are you ready to embark on a journey through the fascinating realms of History, Geography, Maths, Physics, Biology, and Sports? The Ultimate Quiz Book is your ultimate companion, packed with over 500 thought-provoking questions designed to challenge your knowledge and *broaden* your horizons.

Whether you're a seasoned quiz enthusiast or just looking for a fun way to learn something new, this book has something for everyone. Test your wits with questions that cover the span of human knowledge, from ancient civilizations and natural wonders to scientific discoveries and athletic achievements.

Inside, you'll find:

- Engaging Multiple-Choice Questions: Sharpen your skills with a wide range of questions that cater to all difficulty levels.

- Intriguing Facts and Trivia: Discover fascinating titbits that will surprise and delight you at every turn.

- Answers Section: Easily check your answers and learn from any mistakes with a dedicated answers section at the back.

Perfect for individual learning, family fun, or group competitions, The Ultimate Quiz Book is your go-to resource for hours of educational entertainment. Whether you're preparing for a quiz night, a competitive exam, aptitude test or simply satisfying your curiosity, this book ensures you'll never run out of knowledge to explore.